Thinking Critically to Solve Problems: Values and Finite Mathematical Thinking

By Jacci White, Scott White and Siamack Bondari

Saint Leo University

Thinking Critically to Solve Problems: Values and Mathematical Thinking

Preface

Politicians resign in disgrace. Corporate executives are sent to prison. School teachers are arrested for having sex with underage students. Whatever happened to 'common sense'? Shouldn't people just know better? Do Americans still possess a core set of values? You are not alone in considering such questions. Many colleges and universities continue to struggle with them (see http://www.collegevalues.org/). Even if educators could agree on a single set of values, they couldn't be bestowed upon you like a college diploma, with a handshake and a walk across the stage. That said, we believe that an honest intellectual examination of choices and outcomes can bring into focus the set of values by which you already live. Maybe they're working for you. Maybe they're not. This text is not about indoctrination. It's about choice. It's about giving strength to that little thing called intuition that raises a warning flag just before we make poor choices. We've chosen a mathematics course at a small liberal arts college as our setting, but the approach is just as relevant in any course at any institution.

Congratulations! The fact that you are reading this means that you have chosen a different kind of college education. It's not just different, but more comprehensive, more challenging and yes, more rewarding. Let's take a closer look at that 'rewarding' part. Rewards come in many kinds: a great social life, steadfast friends, a high paying job, a big house, a new car, a loving spouse, maybe children. While I do hope that these are in your future, I want you to consider rewards of another kind: self-esteem, confidence, respect, integrity, compassion. These are the essence of a life well lived, and that's where we believe Saint Leo can help. But there is a catch: it's all up to you. You must choose a path. You must take the steps. You must ask the questions. You must find the answers. Do this and your college experience will change your life. You will be transformed, empowered, educated. Not only this, but throughout your life you will transform, empower and educate others.

Contents

Introduction

At Saint Leo University, Benedictine heritage embraces the values of excellence, community, respect, personal development, responsible stewardship, and integrity. But what does that really mean? What does it mean for Mathematics? We hope this brief supplement to your text will help you to answer those questions.

Mission

Saint Leo University is a Catholic, liberal arts-based university serving people of all faiths.

Rooted in the 1,500-year-old Benedictine tradition, the university seeks balanced growth in mind, body, and spirit for all members of its community. At University Campus, at education centers, and through the Center for Online Learning, Saint Leo University offers a practical, effective model for life and leadership in a challenging world; a model based on a steadfast moral consciousness that recognizes the dignity, value, and gifts of all people.

Values

Excellence

Saint Leo University is an educational enterprise. All of us, individually and collectively, work hard to ensure that our students develop the character, learn the skills, and assimilate the knowledge essential to become morally responsible leaders. The success of our University depends upon a conscientious commitment to our mission, vision, and goals.

Community

Saint Leo University develops hospitable Christian learning communities everywhere we serve. We foster a spirit of belonging, unity, and interdependence based on mutual trust and respect to create socially responsible environments that challenge all of us to listen, to learn, to change, and to serve.

Respect

We value all individuals' unique talents, respect their dignity and strive to foster their commitment to excellence in our work. Our community's strength depends on the unity and diversity of our people, on the free exchange of ideas, and on learning, living and working harmoniously.

Personal Development

Saint Leo University stresses the development of every person's mind, spirit, and body for a balanced life. All members of the Saint Leo University community must demonstrate their commitment to personal development in order to strengthen the character of our community.

Responsible Stewardship

We foster a spirit of service to employ our resources to university and community development. We must be resourceful. We must optimize and apply all of the resources of our community to fulfill Saint Leo University's mission and goals.

Integrity

The commitment of Saint Leo University to excellence demands that its members live its mission and deliver on its promise. The faculty, staff, and students pledge to be honest, just and consistent in word and deed.

Saint Leo University Mission and Values retrieved March 22, 2015 from **http://online.saintleo.edu/about-us/mission-values.aspx**

Educational and Learning Goals

1. We expect students to demonstrate intellectual growth:
   - Think critically and independently
   - Make informed decisions
   - Commit to life-long learning
   - Engage in problem-solving
   - Exercise reasoned judgment
   - Develop quantitative skills
   - Learn experientially
   - Understand how living things and physical systems operate
   - Prepare for graduate study
2. We expect students to demonstrate effective communication skills:
   - Speak thoughtfully and respectfully
   - Listen carefully
   - Read critically
   - Write clearly
   - Present information well
3. We expect students to demonstrate deepened spiritual values:

3

- Understand Catholic and Benedictine values and traditions
- Commit to act in concert with one's values
- Respect differences in belief systems and values
- Show compassion and empathy
- Understand the relationships among humans, living things, the universe and God
- Balance one's life
4. We expect students to respond aesthetically:
   - Appreciate the beauty and balance in nature
   - Develop creativity
   - Demonstrate sensitivity
   - Visualize creative potential
5. We expect students to prepare for an occupation:
   - Strive for excellence
   - Develop an international perspective
   - Become competent in: managing people/tasks, responding to change, planning innovation, collaborating, applying technology, acting fiscally responsible
6. We expect students to demonstrate social responsibility:
   - Act with integrity
   - Exercise personal responsibility
   - Respect all living things
   - Work for diversity both locally and globally
   - Build community
   - Commit to resource stewardship
7. We expect students to demonstrate personal growth and development:
   - Develop self-understanding
   - Learn to manage self
   - Deal with ambiguity
   - Exercise flexibility
   - Strengthen confidence and self-esteem

- Learn persistence
- Care for self and physical and spiritual well-being
- Develop leadership
- Foster a work ethic
8. We expect students to demonstrate effective interpersonal skills:
   - Value successful relationships
   - Participate effectively in group work
   - Cooperate
   - Engage in philanthropy
   - Volunteer

Saint Leo University. Inner - Mission and Goals. Retrieved March 22., 2015 from http://www.saintleo.edu/resources/academic-catalogs-schedules-calendars/educational-learning-goals.aspx

Writing Across the Curriculum using Critical Thinking and Values:

1. How can the mission and educational goals affect your approach to this course?

2. How do you feel the mission and educational goals can affect the behavior of your Math instructor?

3. How might the mission and the educational goals make this Mathematics course different from its counterpart at another university?

4. What does the value of integrity have to do with your role as a student in this class? The role of your instructor?

5. Which of the other values should be a part of your experience in this class? How so?

Explorations

In the News: Introduce Yourself:

Please review a newspaper and find an article that is meaningful to your childhood. Use the article to introduce yourself to the class.

Team Values Discussion:

1. Take some time to write your own personal definitions of the six values listed above. Once you have your definitions, break into small groups (or online teams) and share your definitions. As a group, you are to construct a single definition for each value. Having done so, come up with one example of how a college student might experience/apply each value.

2. Write a description of a personal goal that you aim to achieve in this Mathematics class and how a team could help you achieve it.

3. Write a paragraph describing the educational goal you feel is the most important for this level of mathematics and explain why? Share with your team and choose 1 or 2 goals that the team agree are most important.

4. Choose one or more educational goals that you expect to achieve in this mathematics class and write an explanation of how you expect to achieve that goal and how a team might help.

Chapter 1: Set Theory

In this section we will look at ways to combine items into groups and the common mathematical terminology for those associations. The groups we will use are called sets and there are three ways to express a set.

Three methods for describing a set:
- Word description is one method. It is just that, a description of the set in words. The trick is to describe the set specifically enough that every element is included and no extra elements could be mistakenly included.
- Roster notation is another method. This method uses braces to enclose the elements of the set, separated by commas.
- Set builder notation is the last method we will use. For set builder notation, you enclose the set in braces, but begin with the symbols {x: x ∈ and then write the word description before closing the braces}. The symbols x: x ∈ are read "all items x such that x is an element of" the set that is described.

Example:
- Word description: The set of even natural numbers less than or equal to 9.
- Roster notation: {2, 4, 6, 8}

- Set Builder notation: {x: x ∈ the set of even natural numbers less than or equal to 9} another way this can be expressed is {x: x ∈ N and x ≤ 9} where N represents the set of natural numbers.

The **Cardinal number** of a set is the number of distinct elements in the set. If you count each element of a set without counting any like items more than once, that is the cardinal number.

Example:
The Cardinal number, n(A), of the set A= {2, 4, 6, 8} is n(A) = 4 because there are four unique elements in the set.

The Cardinal number of the set B = {2, 4, 4, 6, 8, 8} is n(B) = 4 because there are four unique elements since we do not count repetitions of elements.

Equal sets have identical distinct elements while **equivalent sets** only need to have the same Cardinal number. The order of the elements does not matter.

Example:
Equal: {3, 2, 5} = {5, 2, 3} and {A, N, T, E, A, T, E, R} = {A, N, T, E, R}

Equivalent: {1, 2, 3} = {A, B, C} and {A, N, T, E, A, T, E, R} = (1, 3, 5, 7, 9}
Neither equal nor equivalent: {1, 2, 3} and {m, a, t, h}

A **subset** is a set of elements that all come from the same set. A **proper subset** is a set of elements that all come from the same set and the original set has at least one additional element that is not in the proper subset. Any set with n elements will have 2^n subsets and $2^n - 1$ proper subsets.

Example:
How many subsets and proper subsets does the set A = {a, e, i} have and what are those sets?

Subsets: n = 3 so 2^3 = 8 and they include: { }, {a}, {e}, {i}. {a,e}, {a,i}, {e,i}, {a,e,i}

Proper subsets: n = 3 so $2^3 - 1$ = 7 and they include: { }, {a}, {e}, {i}. {a,e}, {a,i}, {e,i}

The **compliment** of a set is the set that contains all elements that are not in the original set. There are many notations for compliment include: $\overline{A}$, A' and $\sim A$. I will use A' in this book. The **universal set** implies all possible values.

The **intersection** of two sets is the set that contains all elements from the universal set that are in common between the two sets, without repeating the elements. A common symbol for the intersection of sets A and B is $A \cap B$. The **union** of two sets is the set that contains all elements combined from both sets with each distinct element listed only one time. A common notation for the union of two sets is $A \cup B$

Example:
Given the universal set of all Real Numbers less than or equal to 10 with set A = {2, 4, 6, 8} and set B = {1, 2, 3, 4, 5} find:

1. Complement, A': Since A contains the elements 2, 4, 6, and 8, the complement of A will contain the remaining elements in the Universal set of 1, 3, 5, 7, 9 and 10 so A' = {1, 3, 5, 7, 9, 10}.

2. The union of A and B: This contains all elements in A or B or both without repeating any elements so A ∪ B = {1, 2, 3, 4, 5, 6, 8}.

3. The intersection of A and B: This contains all elements that are in both A and B so $A \cap B$ = {2, 4}.

The operations of complement, intersection, and union can be combined in set operations. The operations follow the order of operations so that you perform any operations in parentheses or brackets first, complement comes next, and then perform the set operations from left to right.

Example:
Given the Universal set of all Real numbers less than or equal to 10 with set A = {2, 4, 6, 8} set B = {1, 2, 3, 4, 5} and set C = {3, 4, 5, 6, 7} find:

1. $A \cap B$ ∪ C: Since $A \cap B$= {2, 4} we want to combine this result with set C without repeating any elements. $A \cap B$ ∪ C = {2, 3, 4, 5, 6, 7}.

2. $(A \cap B)'$: This is the complement of the intersection of A and B. That means we find the intersection first, since it is inside the parentheses, then we will find the complement of that result. Since $A \cap B$ = {2, 4}, the complement will be everything else. $(A \cap B)'$ = {1, 3, 5, 6, 7, 8, 9, 10}.

3. A U B U C′ : Working from left to right we will find the union of A and B first A U B = {1, 2, 3, 4, 5, 6, 8} and the complement if C is everything not in C so C′ = {1, 2, 8, 9, 10} so we are ready to find the union of our two results A U B U C′ = {1, 2, 3, 4, 5, 6, 8, 9, 10}.

A **Venn Diagram** is a tool used to visualize sets. Each set is represented by a circle and the elements of the set are inside the circle. Any elements that are not in the set would be on the outside of the circle. When you have two or more sets, you draw them as intersecting circles. Any elements that are in the intersection of the two sets are listed in the intersection of the two circles. Each element should only be listed once within a Venn Diagram. If an element belongs in all the sets, then it is listed in the area where all the sets overlap, so that it only needs to be listed one time.

Example:
Given the Universal set of all Real numbers less than or equal to 10 with set A = {2, 4, 6, 8} set B = {1, 2, 3, 4, 5} and set C = {3, 4, 5, 6, 7} find:

1. The Venn Diagram for A and A'

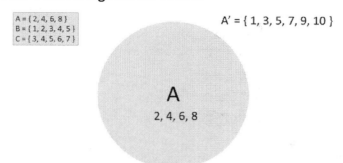

A = { 2, 4, 6, 8 }
B = { 1, 2, 3, 4, 5 }
C = { 3, 4, 5, 6, 7 }

$A' = \{ 1, 3, 5, 7, 9, 10 \}$

A
2, 4, 6, 8

2. The Venn Diagram for $A \cap B$

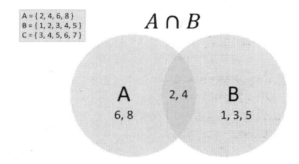

A = { 2, 4, 6, 8 }
B = { 1, 2, 3, 4, 5 }
C = { 3, 4, 5, 6, 7 }

$A \cap B$

A
6, 8

2, 4

B
1, 3, 5

3. The Venn Diagram for $A \cap B \cap$

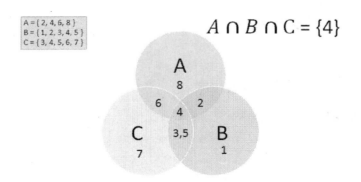

A = { 2, 4, 6, 8 }
B = { 1, 2, 3, 4, 5 }
C = { 3, 4, 5, 6, 7 }

$A \cap B \cap C = \{4\}$

A
8

6   2
4

C   3,5   B
7   1

Explorations

In the News: US and Our World

Check the New York Times, or other national newspaper, to find an article that illustrates some relationship between the United States and another part of the World. Draw a Venn diagram for two sets showing the US as one set and the other part of the World as the second set. Make sure you correctly fill in the diagram to show relevant information included in both sets, each individual set, and neither set. Include all elements that are relevant to the article content.

DON'T FORGET TO REFERENCE!
Here is an acceptable format for referencing your article:

Author (Date). Title of Article. Newspaper Title. Date of retrieval from Name of Database. For Web Sources, give a URL.

For instance:

Troxler, H. (1997, December 31). From Coe to "just say go": Blame it on El Nino. *St. Petersburg Times*. Retrieved March 16, 1998 from ProQuest database.

Athletics & Academics

Take a look at some sets that are used at Universities such as the degree programs or the sports teams.

1. Choose one of the Schools at SLU (Arts and Sciences, School of Business, Education and Social Services, etc.), and either Graduate or Undergraduate. Create a Venn diagram using the set for the school you selected and a set representing Graduate or Undergraduate. Fill in the Venn diagram as completely as possible from the lists of degree programs that you found.

2. Create a Venn diagram choosing two of the following sets: indoor sports, court sports, sports played on a field, pool sports, sports played with a ball, sports played with a racquet or club, men's sport, women's sport, or sport played with a ball that is usually primarily white. Fill in the Venn Diagram using the sports played at a school of your choice.

Values Discussion

1. How can the value of respect be similar between team members in sports and in the classroom?

2. How might it be different?

3. How can embracing the value of respect enhance a classroom environment?

4. How might a classroom environment break down if the value of respect is not embraced by the students?

5. How might a classroom environment break down if the value of respect is not embraced by the teacher?

6. What other values are important if a team is going to be successful?

Writing Across the Curriculum

1. Describe the three methods used to represent a set. Give an example of a set represented by each method.

2. In a given a given exercise, a universal set is not specified, but we know that Hillary Clinton is a member of the universal set. Describe 5 different possible universal sets of which Hillary Clinton is a member. Write a description of one set that includes all the universal sets you came up with.

3. Explain the difference between a subset and a proper subset. Can a set be a proper subset of itself?

4. Describe what is meant by a universal set. Provide an example. Is it possible to find a set's complement if a universal set is not given? Explain your answer.

5. If a set has 127 proper subsets, how many elements are there in the set?

6. Describe the Venn diagram for two equal sets. How does this diagram illustrate that the sets are equal?

7. Describe the Venn diagram for proper subsets. How does this diagram illustrate that the elements of one set are also in the second set?

15

Chapter 2: Linear Equations

To **evaluate** an algebraic expression means to substitute the given values for the variables in the expression and then simplify the result.

Example:
1. Evaluate $f(x) = 2x^2-3x+1$ for $x = -2$

Solution:
First substitute the value -2 in wherever there is an x in the expression.
$f(-2) = 2(-2)^2 - 3(-2) + 1$
$f(-2) = 2(4) + 6 + 1$
$f(-2) = 8 + 6 + 1$
$f(-2) = 15$

2. Evaluate $f(x, y) = -2(x+y) - 3x + 2y$ for $x = 1$ and $y = -2$

Solution:
First substitute the value 1 in wherever there is an x and the value of -2 in wherever there is a y in the expression to get:
$f(1, -2) = -2((1) + (-2)) - 3(1) + 2(-2)$
$f(1, -2) = -2(-1) - 3 + -4$
$f(1, -2) = 2 - 3 + -4$
$f(1, -2) = 2 + -3 + -4 = -5$
$f(1, -2) = -5$

To *simplify* an algebraic expression, follow the order of operations. Remember that you can only combine like terms if the variable part of the term is exactly the same. Like terms have the same variables raised to the same powers. Then combine like terms by combining the coefficient in front of the variable part.

Example:
 1. Simplify $(x-1)^2 + 2x^2 - 3x + 1$:

Solution:
Following the order of operations we would start with the parentheses. However, since the terms are not the same, we cannot combine the x and the 1 so we now go to exponents and multiply:
$(x-1)(x-1)$ to get $x^2 - 2x + 1$.

Our expression now becomes $x^2 - 2x + 1 + 2x^2 - 3x + 1$ and we are ready to combine like terms by combining the coefficients. In front of the x^2 terms we have a 1 and a 2 to get $3x^2$, in front of the x terms there is a -2 and a -3 for a total of -5x, and finally we can add the constant values of $1 + 1$ to get 2 for a simplified form of $3x^2 - 5x + 2$.

Example:
 1. Simplify $2(x-4) - 3(2x - 3) + 5x - 2$

Solution:
Following the order of operations we will use the distributive property to get rid of the parentheses since the terms inside the parentheses cannot be combined. Our expression becomes:

2x – 8 – 6x + 9 + 5x – 2

Now we can combine the x terms and the constants to get:
2x – 6x + 5x – 8 + 9 – 2 = x – 1.

Translating English phrases and algebraic expressions

Students often have a fear of word problems. If you read problems aloud, they are in words, so a large part of solving word problems is understanding the symbols that represent different words. For example, is and equal are interchangeable. Increased by and combined with would imply addition while decreased by or less than could mean subtraction.

Example:
The English statement: Six times a number increased by seven times the number minus fifteen translates into:

6x + 7x - 15   which can be simplified to:
13x – 15.

The English statement: Five times the sum of a number and two is reduced by three times the difference of the number and one translates into:
5(x+2) – 3(x-1)

This is because five times the sum implies we will start with 5 and multiply it by some parenthesis that contains a sum. The sum inside the parentheses is a number and two or x+2 so we have 5(x+2). To continue, we want to reduce this by 3 times a difference so that is subtracting 3 times parentheses and the parentheses will contain subtraction. Specifically, that subtraction is the difference of a number and 1 or x-1, to get 3(x-1).

5(x+2) − 3(x-1)   which can be simplified to:
5x + 10 − 3x + 3
2x + 13

An **equation** is different from an expression because it has an equal sign and can therefore we can solve for the variable. You solve an equation for a variable by getting all terms with that variable on one side, all other terms onto the other side, and then divide both sides by the coefficient in front of the variable you are solving for. There are several rules that can be used to isolate the variable on one side of the equation.

Addition Property of Equality: For all real numbers, x, y, and z:
If x = y then x + z = y + z

This means you may add any number or term as long as you add the same thing to both sides of the equation. The sides are separated by the equal sign so that means you add the same thing on both sides of the equal sign. You may also subtract any term you want as long as you subtract it from both sides because subtraction is the same thing as adding the opposite value, the negative value. We will use this idea to get the variable term isolated on one side of the equation by subtracting it from any side where we do not want the variable. Then, we will use this idea again to get all other terms to the other side of the = sign by subtracting any terms on the variable side of the equation if they do not contain the variable.

Multiplication Property of Equality: For all real numbers, x, y, and z:
If $x = y$ then $x*z = y*z$

This means you may multiple both sides of an equation by the same number or expression. You can also divide both sides by the same number or expression since division is the same a multiplying by the inverse. Therefore, your last step will be to divide by the coefficient in front of the variable.

Example:
1. Solve the following equation for x: $2x - 3 = 5x + 6$

Solution:
In this case I will start by subtracting 2x from both sides to get all the x terms on the right.
$2x - 3 - 2x = 5x + 6 - 2x$
$-3 = 3x + 6$

Now subtract 6 from both sides to get the x term isolated on the right side of the equation with all other terms on the left. Notice I am subtracting 6 because the 6 in the equation is being added so I need to subtract 6 to cancel out the 6 in the equation:

-3 - 6 = 3x + 6 – 6
-9 = 3x.

Finish by dividing by the coefficient in front of x, so divide both sides by 3 to get:
-9/3 = 3x/3
x = -3

One nice thing about solving Algebraic equations is that you can always check your answer. If you substitute x = -3 into the equation we started with, it should simplify to a true statement.
2(-3) – 3 = 5(-3) + 6
-6 – 3 = -15 + 6
-9 = -9

This is a true statement so we have a correct solution.

Some equations must be simplified before you can begin isolating the variable. For example, if there are any fractions, you should multiply both sides by the least common denominator to eliminate the fractions simplifying the arithmetic and eliminating the fractions. Also, you should multiply out parentheses whenever possible. And, finally, you should combine like terms on each side before you start moving terms between sides.

Example:
    1. Solve for x: $3(x - 2) + 3 = 8 - 2(3x+1)$

Solution:

Since this equation has parentheses, I will start by simplifying the equation individually on each side of the equation to get:

$3x - 6 + 3 = 8 - 6x - 2$

$3x - 3 = 6 - 6x$

Now you can either subtract 3x from both sides to get the x terms on the right side of the equation or else add 6x to both sides to get the x terms on the left side of the equation. I choose to add 6x to both sides of the equation to get:

$3x - 3 + 6x = 6 - 6x + 6x$

$9x - 3 = 6.$

Now add 3 to both sides to move the non x term onto the other side of the equation.

$9x - 3 + 3 = 6 + 3$

$9x = 9.$

Lastly, divide by the coefficient in front of x to get x by itself:

$9x/9 = 9/9$

$x = 1$

Example:

1. Solve for x: $\dfrac{3x}{2} + 1 = \dfrac{4}{3} - 2x$

Solution:

We can simplify this equation by eliminating the fractions. We multiply both sides of the equation by 6 since that is the least common denominator.

$$6\left(\dfrac{3x}{2} + 1\right) = 6\left(\dfrac{4}{3} - 2x\right)$$

$$6 * \dfrac{3x}{2} + 6 * 1 = 6 * \dfrac{4}{3} - 6 * 2x$$

9x + 6 = 8 − 12x.

Now that the equation is simplified, the fractions have been eliminated, we are ready to get all the variable terms on one side. I will add 12x to both sides to get the x terms on the left:

9x + 6 + 12x = 8 − 12x + 12x

21x + 6 = 8.

Next, I will subtract 6 from both sides to isolate the x term on the left:

21x + 6 − 6 = 8 − 6

21x = 2

Lastly, I divide both sides by 21 to get x by itself:

$$\dfrac{21x}{21} = \dfrac{2}{21}$$

$$x = \dfrac{2}{21}.$$

Example:

1. A University currently has space to have a total of 23,400 students in a class per year (52 classrooms, 15 classes each, 30 students per class). They renovate old buildings and offices to add space for 450 more students in classes per year (1 classroom, 15 classes, 30 students per class). The current enrollment capacity is 21,900 (4,380 students with 5 classes each) class seats and growing at 600 class seats needed per year (120 new students with 5 classes each). In how many years will the student capacity exceed the classroom capacity?

Solution:

Let x represent the number of years, since that is what we are looking for. The equation to represent the classroom space is the 23,400 current seats in classes plus the 450 new seats times x or:

Total available seats: $24,400 + 450x$

Total student seats needed is 21,900 for current students plus 600 times x for the 600 new ones needed each year or:

Total needed seats: $21,900 + 600x$

We want to know in how many years, x, will the number of available seats equal the number of needed seats, so we set these two equal to each other.

Total available seats = Total needed seats
$24,400 + 450x = 21,900 + 600x$

Solving for x:
Subtract 450x to move the x terms to the right side of the equation:
$24{,}400 + 450x - 450x = 21{,}900 + 600x - 450x$
$24{,}400 = 21{,}900 + 150x$

Subtract 21,900 to isolate the x terms on the right:
$24{,}400 - 21{,}900 = 21{,}900 + 150x - 21{,}900$
$1500 = 150x$

Divide both sides by 150 to solve for x:
$1500/150 = 150x/150$

$10 = x$

In 10 years the seats needed will equal the seats available and there will be a shortage of seats the following, or 11th year.

The same process is used to **solve inequalities** like $<$, $\leq$, $\geq$, or $>$ as you use for equalities, with one addition. Whenever you multiply or divide both sides of the inequality by a negative number, the sense of the inequality is reversed (the inequality symbol is flipped).

Example:
1. Solve the following equation for x: $2x - 3 \leq 5x + 6$

Solution:
In this case I will start by subtracting 2x from both sides to get all the x terms on the right.
$2x - 3 - 2x \leq 5x + 6 - 2x$

$-3 \leq 3x + 6$

Next subtract 6 from both sides to get the x term isolated on the right side of the equation with all other terms on the left. Notice I am subtracting 6 because the 6 in the equation is being added so I need to subtract 6 to cancel out the 6 in the equation:
$-3 - 6 \leq 3x + 6 - 6$
$-9 \leq 3x$

Finish by dividing by the coefficient in front of x, so divide both sides by 3 to get:
$-9/3 \leq 3x/3$
$x \leq -3$

Notice that the inequality did not reverse since we divided by a positive 3 not a negative number.

Example:
   2. Solve $3(x - 2) + 3 \geq 8 - 2(3x+1)$ for x.

Solution:
   Since this equation has parentheses, I will start by simplifying the equation individually on each side of the equation to get:
$3x - 6 + 3 \geq 8 - 6x - 2$
$3x - 3 \geq 6 - 6x$

Now I will subtract 3x from both sides to get:
$3x - 3 - 3x \geq 6 - 6x - 3x$
$-3 \geq 6 - 9x$

I will subtract 6 from both sides to get:
$-3 - 6 \geq 6 - 9x - 6$

-9 $\geq$ -9x

I need to divide by -9 to isolate for x. Since it is a negative number, I will also have to reverse my inequality to get:

-9/-9 $\geq$ -9x/-9
1 $\geq$ x

Example:
3. A car can be rented for $60 per day plus $0.20 cents per mile or $78 per day with unlimited miles. How far would you need to drive before the cost of unlimited miles is a better deal?

Solution:
For this problem, the cost when paying by the mile would be given by: 60 + 0.2x

where x is the number of miles driven. We want to know when this is less than or equal to $78 to determine the point when you should switch to the higher base cost. Our equation looks like:
60 + 0.2x $\leq$ 78

Subtracting 60 from both sides we get:
0.2x $\leq$ 18

Now we divide by 0.2 and since it is positive, the inequality symbol remains the same:
0.2x/0.2 $\leq$ 18/0.2
x $\leq$ 90

That means you should rent at the $60 rate if you plan to drive less than 90 miles per day and at the $78 rate if you plan to drive more than 90 miles per day. The rates are equal if you drive exactly 90 miles per day.

Example:
4. A taxi charges a flat rate of $2.50 plus $0.50 for the first fifth of a mile and $0.50 for each additional fifth of one mile. How far can you travel toward your destination if you have $12.00?

Solution:
The unknown in this problem is the total number of fifths of a mile you can travel so we let x equal then number of fifths of a mile. That means we can create an equation for the cost of the trip. Since the flat charge is $2.50, you start with that and add 50 cents per fifth of a mile, or 0.50 * x and the charges must max out at $12 so the charges must be less than or equal to 12.

$2.5 + 0.50 * (x) \leq 12$

To solve this equation first get the x term by itself by subtracting 2.5 from both sides of the inequality:
$2.5 - 2.5 + 0.50 * (x) \leq 12 - 2.5$
$0.5x \leq 9.5$

Next, divide both sides by 0.5 to isolate a single x term. Since 0.5 is positive, the direction of the inequality does not change, resulting in the number of fifth of a mile being less than or equal to 19 that you can travel for $12.00.
$0.5x/0.5 \leq 9.5/0.5$
$x \leq 19$

Since x represents fifths of a mile, to determine the total number of miles, we divide 19 by 5 to see how many miles we can drive resulting in 19/5 = 3.8 miles or less can be traveled for $12.00.

Explorations

In the News: Terrific Tuesdays

The New York Times has a special Science section every Tuesday. From the science section, find an article that uses a mathematical concept from this class and illustrate how the mathematical concept is used.

DON'T FORGET TO REFERENCE! See *Section 1* for an example of how.

Writing Across the Curriculum

1. Write a report explaining why we use m to represent the slope of a line and explaining why we call the x-y coordinate system the Cartesian coordinate system.

2. Find a graph of an equation in the media and explain how it is used and how the graph was obtained.

3. Why can the solution to an inequality in one variable not be a single number?

4. Research a mathematician who has an Algebra formula named after him/her and explain why the formula is named after that mathematician as well as giving a brief biography of the mathematician.

5. Write a report explaining how an equation and the graph of the equation are similar and how they are different. Include the benefits and disadvantages of each.

Critical Calculations

1. Find an equation that has three or more variables and solve it for all three variables.

Critical Thinking for Reflection with Values

1. Do you have to respect the mathematician to respect the mathematics he is credited with?

2. How did the mathematician you explored have, or not have, personal Integrity?

3. How is studying Algebra important for Personal Development? Is that different from studying other forms of mathematics?

Chapter 3: Functions, Graphs, and Systems of Equations

We just saw how to solve linear equations in one variable. We also saw how to evaluate a linear expression when there is more than one variable. Now we will look at solutions to a linear equation in two variables and progress on to find a solution that satisfies multiple linear equations, called a system of equations.

This is a good time to introduce function notation. An equation is a function if for any value of the independent variable, there is at most one value of the dependent variable that satisfies the equation. That means if you graphed a function, a vertical line would not pass through it more than one time (the Vertical Line test) because there would be at most one y-value for each x-value, giving you at most one point for each value of x. If an equation satisfies the conditions of a function, then it can be written in function notation, replacing the y with $f(x)$ (or occasionally $g(x)$, $h(x)$ or some other letter to represent the particular function being considered, but most often it will be $f(x)$). This notation allows us to express the independent variable we might be using to find the dependent variable.

Example:
1. If we wanted to evaluation the function $y = 2x + 1$ for the value of $x = 3$, using function notation, the original equation would look like:
$f(x) = 2x + 1$

   To evaluate the equation it would look like:
$f(3) = 2x + 1$

Then we can evaluate the function by substituting 3 for x to get:

f(3) = 2(3) + 1

f(3) = 6 + 1

f(3) = 7

2. Evaluate f(x) = 2 – 4(x) for x = -1:

f(-1) = 2 – 4(-1)

f(-1) = 2 + 4

f(-1) = 6

Graphing Functions

To plot a function, you choose values of x or y to substitute into the equation in order to solve for the other variable to get ordered pairs. Then plot two ordered pairs on the coordinate system and draw a line that extends through them. If you choose to let x = 0 and solve for y, then let y = 0 and solve for x, your two points will be the x and y intercepts of your graph that you then extend the line through. However, any two points will work.

When you want to sketch the graph of an inequality, you start by graphing the equation as if it were an equality, using a solid line if it includes the equality such as ≤, ≥, and using a dotted line if it is just <, or >. Once you have that portion of the graph done, then choose any point that is not on the line you drew. You use that point to substitute into the original inequality. If the result is a true statement, then you shade the side of the line that includes the point. If the result is a false statement, you shade the other side of the line.

Example:
1. Sketch the graph of f(x) = 2x + 1

Solution:

To find two ordered pairs to plot the line, we will let x = 0:

f(0) = 2(0) + 1

f(0) = 1 or the point (0,1)

To find the second point, we will let x=1:

f(1) = 2(1) +1

f(1) = 2+1

f(1) = 3 for the point (1,3).

We could have chosen any two value for x, but the arithmetic is simpler when choosing numbers such as 0 and 1. Now plot these two points and then extend a line that extends through these two points.

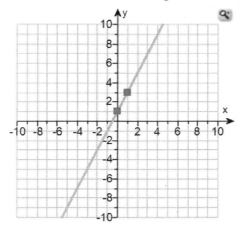

(screen shot from mypearsonlabandmastering.com)

You can also graph an equation using the slope and y-intercept. In that case, you plot the y-intercept and then count rise over run to the next point on the line.

Example:

2.

Graph the linear function using the slope and y-intercept.

$$f(x) = -\frac{3}{4}x + 6$$

Solution:

This equation is in the slope-intercept form:

y = mx + b

Where m is the slope and is the y-intercept. Therefore, y-intercept is 6 so we could start with a point at (0,6) because the x value of the y-intercept is 0, the x-axis. The slope of -3/4, the m term in the equation, means to go down (down because it is negative) 3 units and over to the right 4 units to get to the second point on the graph, (4,3), in order to connect them with a line that extends through the two points.

Alternatively, you can still plot this example using the intercepts or two points as well. If x = 0, then you can solve for:
y= -¾(0) + 6
y = 6 to get the point (0, 6).

Then let y = 0 and solve for x:
0 = -3/4x + 6
-6 = -3/4x
(-4/3)(-6) = x

8 = x for the second point of (8, 0).

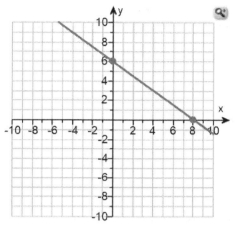

Solving Systems of Equations

We are now ready to solve systems of equations. That means we find solutions that satisfy more than one equation or inequality at the same time. There are several methods for solving a system of linear equations and you can reference a textbook to see the substitution method, or methods that utilize matrices. In this book we will use the elimination method for solving systems of equations.

In order to use the substitution method:

First, put the equations in standard form so the variable terms are on the left in alphabetical order and the constant terms are on the right.

Second, look at the variable terms and determine what constant we should multiply each equation by so that when we add the equations together one of the variables will drop out (the sum will be zero for that variable). Multiply both sides of one equation by this constants.

Third, subtract the two equations. This will leave us with an equation in only one variable.

Forth, we can solve this single variable equation by isolating the variable on one side of our equation just as we did in the last chapter.

Fifth, once we have a solution for one variable, we can substitute that value into either original equation to solve for the other variable resulting in the ordered pair that represents our complete solution.

Example:
1. Solve the following system of equations:
$$3x - y = 10$$
$$2x + y = 5$$

Solution:
Notice that the coefficients of y in these equations are already opposite values so when we add the two equations the y values drop out:

$$3x - y = 10$$
$$\underline{2x + y = 5}$$
$$5x + 0y = 15$$
$$5x = 15$$
$$5x/5 = 15/5$$
$$x = 3$$

That means x=3 is half of our solution. Now we can substitute x=3 into either of the original equations and get the same value of y since that value is the second half of our solution. I will choose to substitute x into the first equation to get:

$3x - y = 10$
$3(3) + y = 10$
$9 + y = 10$
$9 - 9 + y = 10 - 9$
$y = 1$

That means the solution set is the ordered pair (3, 1). If you substitute this result into the equations, they will both be true statements because this is the point where the two equations intersect, or the solution they have in common.

Example:

2. Solve the system:
$2x + 3y = -1$
$3x + 6y = -3$

Solution:

For this system, I have decided to eliminate the y variable again because I can get opposite coefficients in one step by multiplying the top equation by -2:
$-2(2x + 3y) = -2(-1)$
$-4x - 6y = 2$

Now I am ready to add the two equations:
$-4x - 6y = 2$
$\underline{3x + 6y = -3}$
$-x + 0y = -1$
$-x = -1$
$x = 1$

The result of x=1 for the first half of the solution. Substituting x=1 into the first equation I have:
$2x + 3y = -1$
$2(1) + 3y = -1$ Substitute x = 1

2 + 3y = -1
2 – 2 + 3y = -1 -2   Subtract 2 from both sides
3y = -3
3y/3 = -3/3   Divide both sides by 3
y = -1

The result is y = -1 for a solution of x = 1 and y = -1 that can be expressed as the ordered pair (1, -1).

Example:
3. Solve the system:
9x – 3y = 4
3x – y = 2

Solution:
For this system, I am going to multiply the second equation by -3 so the x values will drop out when I add them. My resulting system is:
-3(3x + y) = -3(2)
-9x -3y = -6

9x – 3y = 4   Add the two equations
-9x -3y = -6
——————————
0x – 0y = -2
0 = -2

When I add them and I end up with 0 = -2. Notice both variables dropped out and I have a false statement, because 0 is never equal to -2. This means these lines are parallel and never cross so they have no points in common, or no common solution.

Example:
4. Solve the system:
9x – 3y = -6
3x – y = -2

Solution:

For this system, I will again multiply the second equation by -3 to eliminate the x values when I add the equations:

$3x - y = -2$

$-3(3x - y) = -3(-2)$

$-9y - 3y = 6$

$9x - 3y = -6$

$\underline{-9x + 3y = 6}$

$0x + 0y = 0$

$0 = 0$

Notice when I add the equations now I get $0 = 0$. Since this is always a true statement, there are infinitely many solutions to this system. That does not mean all points are solutions (notice (0, 0) is not a solution to this system), it means that these are two equations for the same line, or that the lines are on top of each other. That means every point on either line is a solution. We can write our solution set as:

{All Real (x,y) such that $3x - y = -2$}.

This means all real points that are on the line are solutions to the system.

Solving systems of linear inequalities:

The solution to inequalities in two variables is infinite and is represented by all points on the shaded region of our graph of the inequality (and including the values on the line itself when the inequality includes the equality). The same is true for systems of inequalities, the solutions can only be shown as a shaded region of a graph. As a result, solving systems of linear inequalities in two variables means we sketch the graph of each inequality on the same coordinate system. The solution to the system is the area where the shaded regions intersect.

Example:
1. Solve the following system of inequalities:
    $y > -6x - 8$
    $y < -x + 3$

Solution:
We first treat each inequality as an equality for the sake of finding two points that fall on the line. The first step is to graph each line. For the first inequality:
First let x=-1:
$y=-6(-1)-8$
$y = 6-8$
$y = -2$ for the point (-1,-2).

Then let x=0:
$y = -6(0) - 8$
$y = -8$ for the point (0,-8)

Now plot the line with these two points but choose the dotted line feature since it is > but not equal to.

Once you have this line sketched, we need to determine which side of the line makes the inequality true. We choose a test point such as (0, 0) and we substitute this point into our inequality to determine if the statement is then true:
$y > -6x - 8$
$(0) > -6(0) - 8$
$0 > -8$

This is a true statement so we shade the side of this line that contains our test point, (0, 0).

Next we need to graph the other inequality by determining two points that satisfy the inequality if it were an equality.

First, let x = 0:
y = -(0) + 3
y = 3 for the point (0,3)

Then let x = 1:
y = -(1) + 3
y = -1 + 3
y = 2 for the point (1,2)

Now we plot the line thru these two points but making the line itself dotted since this is a less than inequality (not less than or equal to). Lastly, to determine the side of tis inequality that makes the inequality true, we choose a test point, (0,0) will work again:
(0) < -(0) + 3
0 < 3 : True

Since this statement is true, we shade the side of the inequality that includes the test point, (0,0). We can use any point that is not on the line as a test point, using (0,0) tends to make the arithmetic simpler. The resulting graph is below:

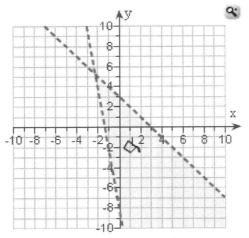

(screen shot from mypearsonlabandmastering.com)

Example:

2.

Graph the solution set of the following system of inequalities.

$$y > 2x + 10$$
$$y < -x - 5$$

Solution:

For this problem start by sketching the first graph. Remember to use a dotted line because the inequality does not contain the equality. Again, pick two points that satisfy the equation, then draw the dotted line.

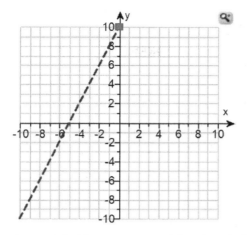

(screen shot from mypearsonlabandmastering.com)

Now sketch the second equation on the same coordinate system.

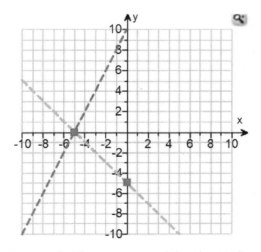

(screen shot from mypearsonlabandmastering.com)

Now we will shade the appropriate section of the coordinate plane. We choose a test point for each equation to see which side of the lines to shade. I choose the point (0,0) because it is clearly not on either line and this point will make the arithmetic relatively simple.

$y > 2x + 10$
$y < -x - 5$

$0 > 0 + 10$ FALSE
$0 < 0 - 5$ FALSE

I will see that both equations are false at (0,0). That means I should shade the side of each line that does not include the point (0,0) because it was false. The final answer is the area where the two regions intersect. In this case that is the left quadrant of the graph.

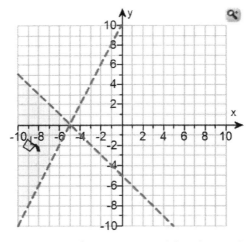

(screen shot from mypearsonlabandmastering.com)

45

Explorations

In the News: Terrific Tuesdays

The New York Times has a special Science section every Tuesday. From the science section, find an article that uses more than one graph on the same axis. Explain what the graphs illustrate together that is not shown when only one graph is present.

DON'T FORGET TO REFERENCE! See *Section 1* for an example of how.

Writing Across the Curriculum

1. Explain how many equations are needed to solved a system that has three unknowns. How about a system with four variables? What is the relationship between the number of variables you are solving for and the number of equations needed to solve the system?

2. Explain when a system of two equations in two variables might not have any solutions. Make sure your explanation includes what the graph of this situation looks like.

3. Explain when a system of two equations with two variables might have infinitely many solutions. Make sure your explanation includes what the graph of this situation looks like.

Teams for Social Justice

1. As a team, find a graph that illustrates population and another that illustrates food production or availability. Make sure the two graphs are for the same population (whatever population you choose). Where do these two graphs intersect? Does that "solution" imply that the population should or should not have sufficient food?

2. Find graphs or equations that illustrate two of either women's wages, overall wages, and/or men's wages over time in the same type of jobs. Analyze the data to see if genders appear to be paid the same amount for the same job. Is experience considered in the data? Find a study comparing men's and women's pay and look at the information used to determine if other variables were held constant within the study so that the results are a true comparison of men and women in the same job with the same education and experience.

Chapter 4: Linear Programming

Linear programming is a fancy name for solving systems of linear equations. In linear programming, the equations represent specific situations where you have a main equation representing something you want to maximize or minimize subject to certain constraints, such as maximizing profit within limits of time and money. The main function that you wish to maximize or minimize is called the "objective function". The limits to this objective function are called constraints.

Linear programming allows us to maximize or minimize the objective function within the limits of the constraints. The method for doing follows:

First, develop the objective function and the system of constraint equations.

Second, we analyze the constraints as a system of inequalities and find the points where the lines cross forming the boundaries of the solution to the system of constraints. For two linear constraints, there will be the following four points to evaluate:
1. The origin
2. The x intercept
3. The y intercept
4. The intersection of the two constraints

Third, we test the objective function at each of these points. Since the objective is a linear function, the maximum or minimum will occur at one of these boundary corners. So we solve for objective function by substituting these corner ordered pairs into the objective function. This will provide us with the minimum or maximum value of the objective function.

The steps for these calculations can be time consuming, so software is often used to solve the system of equations after the equations have been set up. See below the first example for a link to free software to solve these problems as well as instructions of the commands used with the software.

Example:
1. Clothing and food are sent to people in need in a remote area. Each container of food will feed 12 people while each container of clothing holds enough for 5 people. The food containers are 25 cubic feet and weigh 50 pounds while the clothing containers are 5 cubic feet and weigh 20 pounds. The airplane can hold at most 21,000 pounds and 7000 cubic feet. How many containers of food and clothing should be collected and shipped in order to help the most people?

Solution:
Develop the objective function. The variables in this examples are the number of containers of food and containers of clothing. Therefore, We will create the following two variables:
x: containers of food
y: containers of clothes

In this case the object is to serve the maximum number of people. Since food containers serve 12 people and clothing containers serve 5 people, the objective function is:
$z = 12x + 5y$

Develop the constraint equations: You will have two constraint equations, one for weight and one for space.

Weight: Each carton of clothes weighs 20 pounds and each carton of food weighs 50 pounds for the following equation:
50x+20y ≤ 21000

Space: The clothes cartons taking up 5 cubic feet and the food carton taking up 25 cubic feet for the following equation:
25x+5y ≤ 7000

Evaluate the objective function at the four critical points. The four points are:

The origin: (0, 0). This is the simplistic case of doing nothing.

X and y Intercepts: We would look at this equation at all the points of intersection. The intercepts of the weight equation, 25x+5y<=7000, are (280, 0) and (0, 1400). You can find these points by substituting 0 for x, solving for y, then substituting 0 for y, solving for x. The intercepts of the space equation, 50x+20y <=21000, are (420, 0) and (0, 1050). We need to use the smaller of the constraints on the x and y axis, so we will use (280, 0) and (0, 1050).

The fourth point is the intersection of the two constraint equations. To find this point, we solve this by addition as we did in the previous chapter. I will multiply the space equation by -2 and add the two equations.

$$50x + 20y \leq 21000$$
$$\underline{-50x + -10y < -14000}$$
$$0x + 10y = 7000$$

$$y = 700$$
$x = 140$, for a fourth point of (140, 700)

We now have four points to evaluate the objective function:
$$z = 12x + 5y$$

Origin: $z = 0$; no one gets served if we do nothing
X-Intercept: $z = 12*280 = 5040$ people served
Y-Intercept: $z = 5*1050 = 5250$ people served
Intersection: $z = 12*140 + 5*700 = 5180$ people served.

So to serve the maximum number of people, the maximum is at the Y-intercept which implies sending just clothing containers we can serve 5,250 people. This means (0, 1050) is the maximum value so you would collect 0 containers of food and 1050 containers of clothes.

Technology: You might consider downloading QM for Windows from:

http://wps.prenhall.com/bp_render_qam_11/197/50557/12942690.cw/index.html

We can solve this same problem using Open QM for Windows. Open QM for Windows and from the menu titled "modules" select "linear programming" then from the "file" menu select "new" and you will be prompted for the information from this problem. We have two constraints, weight and volume. There are two variables, food and clothing. It is a maximize problem.

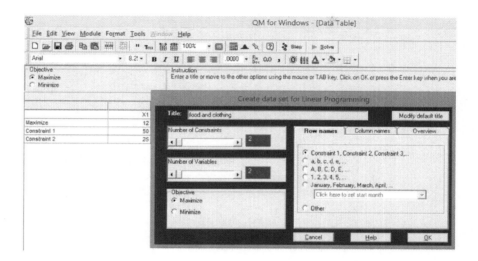

Now we select "ok" and then will enter the objective function coefficients and the constraint coefficients.

Z = 12x+5y

50x+20y <=21000

25x+5y<=7000

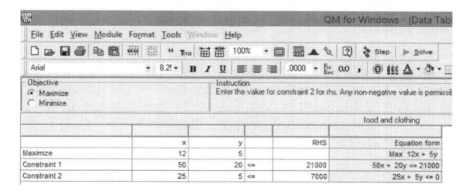

Now press "Solve" to conduct the calculations and see the result of y=1050 and Z=5250 in the final row titled "Solution."

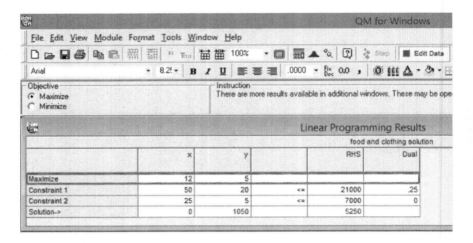

Example:
2. A youth group is holding a performance to raise money for an upcoming mission trip. It is also an outreach to other youth, so every two parents must bring at least one child and there are only 210 seats. Tickets are $10 for adults and $5 for kids. How many parents and students should attend to raise the most money?

Solution:
For this problem, the objective function is total proceeds so Z=10p+5s, where p is the number of parent tickets and s is the number of student tickets since parent tickets are $10 each and student tickets are $5 each.

The constraints are that p+s $\leq$ 210 since the total number of people must be less than or equal to the theater capacity of 210.

The second constraint is that the number of students has to be greater than or equal to half the number of parents there, meaning there cannot be more than twice the number of parents than there are students since every two parents must bring at least 1 child. This gives the final constraint equation of:

p/2 $\leq$ s   or multiplying both sides by 2 gives us

p $\leq$ 2s

p − 2s $\leq$ 0

To solve for the intersection, we will graph both constraints. We graph the capacity constraint p + s = 210 and p-2s = 0.

p+s=210	p	s
	0	210
	100	110
	210	0

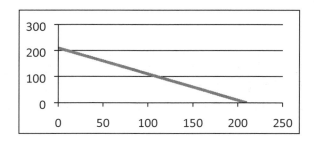

Next graph p-2s=0

p-2s=0	p	s
	0	0
	1	2
	2	4

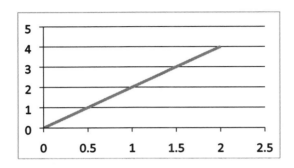

Here they are together:

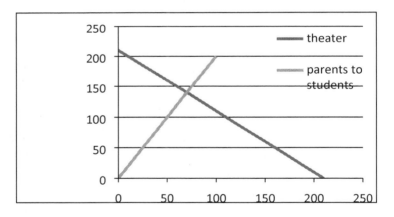

So we see the intersection points are (0,0), (210,0) and where they intersect that we get by solving them as a system.

p + s = 210
2p = s

Since one equation is already in terms of the other, we can use substitution but substituting 2p for s in the first equation.

p + (2p) = 210
3p = 210
p = 70, then, since s = 2p:
s = 140

We now have three points to test (three since the origin and the y-intercept are the same point):
(0, 0)
(210, 0)
(140, 70)

The objective is to maximize ticket revenue:
Z = 10p + 5s
Z(0,0) = 10*0 + 5*0 = 0
Z(210,0) = 10*0 + 5*210 = 1050
Z(140,70) = 10*140 + 5*70 = 1400 + 350 = 1750

Therefore the maximum ticket revenue would result from selling 70 student tickets and 140 adult tickets.

Using QM for Windows I type in my objective function and constraints and you can see the results here.

			QM for Windows
File Edit View Module Format Tools Window Help			
Objective		Instruction	
Maximize Minimize		There are more results available in additional windows. These may be oper	

Linear Programming Results

theater solution

	p	s		RHS	Dual
Maximize	10	5			
Constraint 1	1	1	<=	210	8.3333
Constraint 2	1	-2	<=	0	1.6667
Solution->	140	70		1750	

Explorations

Writing Across the Curriculum

1. Why is linear programming important in business? What are other areas where linear programming is used?

2. Explain the steps in a linear programming problem. Include examples.

3. Why can solutions to systems of inequalities only be shown graphically?

4. Each linear programming model that is feasible has infinitely many solutions. Explain. What is different about the particular solution you search for in a linear programming model?

Critical Calculations

1. Post the steps for solving a linear programming problem using technology.

Critical Thinking for Reflection with Values

1. How can linear programming be used by a business to enhance Responsible Stewardship and respect?

2. We often use technology to help us solve mathematics problems. What might the phrase "respect the technology" mean?

3. In what way does Excellence in business require Respect?

4. Do you have to respect the business man to respect his business? How about to do business with him?

5. If you respect your employees and customers, what are some situations where you might not go with the optimal mathematics solution from a linear programming problem, but adjust the result even though it will likely be less profitable?

Chapter 5: Consumer Mathematics

In this chapter, we will begin with simple problems, but we will quickly progress to more complex calculations. We strongly recommend that you should familiarize yourself with a financial tool such as a financial calculator or Excel to carry out your calculations before they become tedious. You will need to learn how to go back and forth between the decimal and percentage forms of a number because interest rate is often given in percentage form, but is always used in decimal form for calculations. Remember, to convert a percentage to decimal form, you drop the percentage sign and move the decimal two places to the left. Decimal to percentage is the reverse. We start off the chapter with examples of sales tax, discount, percent increase, and percent decrease. For all the examples, you will need to convert numbers from percent to decimal or the other way around.

Example: (Sales Tax)
The Florida state tax rate in 2014 was 6%. On a trip to grocery store, Mrs. Doubtfire purchased $125 worth of groceries. Calculate the sales tax for her purchase. What is the total amount Mrs. Doubtfire spent at the grocery store? Round off your answers to the nearest cent.

Solution:
Sales tax = (tax rate)(purchase)
$$= (6\%)(\$125)$$
$$= (0.06)(\$125) = \$7.50$$

Total amount = purchase + tax
$125+$7.5 = $132.50

Another way to calculate the total amount is:
Purchase * tax
Purchase + (purchase * tax rate)
Purchase * (1 + rate)
($125) (1.06)= $132.50

Example: (Discount)
A graphing calculator was advertised for $92. The calculator is now on sale at 12% off. What is the amount of discount and what is the current sale price for the calculator? Given that the sales tax in the state is 6.5%, calculate the total cost of purchasing the calculator. Round off your answers to the nearest cent.

Solution:
Discount amount = (0.12)($92) = $11.04
Sale Price = $92 - $11.04 = $80.96
Total amount = (1.065)($80.96) or approximately $86.22

Example: (Percentage Increase)
According to the US Census Data, population of the United States increased from 151,325,798 in 1950 to 281,421,908 in 2000. The US population in November 2014 was estimated to be 319,168,000.
a) Calculate the population percentage increase from 1950 to 2000.

b) Calculate the population percentage increase from 2000 to 2014.

Solution:

a) Population percent increase from 1950 to 2000

$$= \frac{Amount\ Increase}{Original\ Amount}$$

$$= \frac{281,421,908 - 151,325,798}{151,325,798} = \frac{130,096,110}{151,325,798} \approx 85.97\%$$

b) Population percent increase from 2000 to 2014

$$= \frac{Increase}{Original\ Amount}$$

$$= \frac{319,168,000 - 281,421,908}{281,421,908} = \frac{37,746,092}{281,421,908} \approx 13.41\%$$

Example: (Percent Decrease)

A Honda civic was purchase at $18,500 in 1998. After ten years and 180,000 miles, the car was sold at $2,000 in 2008. What was the percent decrease in the value of the car from 1998 to 2008?

Solution:

Percent decrease of the car value

$$= \frac{18,500 - 2,000}{18,500} = \frac{16,500}{18,500} \approx 89.19\%$$

Calculating Simple Interest

Simple interest is an example of linear growth, the amount of interest is a percentage of the principal deposit and it remains fixed during each period of time.

The formula for the simple interest is given by

Interest = (Principal)(Rate)(Time) or $I = \mathrm{Pr}t$.

The interest rate r is given in percent and is converted to decimal in calculations.

The future value A for a simple interest is given by

Future value for simple interest $A = P + I = P + \mathrm{Pr}t$: $A = P(1 + rt)$,

Where $P =$ the principal amount also known as the present value, $I =$ the interest, $r =$ the interest rate, and $t =$ time in years.

Here are a few examples of simple interest.

Example: (Simple Interest)
You deposit $2,000 in a saving account in the Dream Bank at the simple interest rate of 1.75% for 2 years. What will be the interest at the end of the period and how much money will you have at the bank at that time? Round off your answers to the nearest cent.

Solution:

Interest = $I = \Pr t = (\$2{,}000)(0.0175)(2) = \$70.00$

Future value = $A = P + I = \$2{,}000 + \$70 = \$2{,}070.00$

Example: (Simple Interest Rate)

You borrow $1,200 from a friend. After 9 months she asks for $1,500. Assuming that she used simple interest to calculate the amount of interest, what was the interest rate? Round off your final answer to two decimals.

Solution:

$I = \$1{,}500 - \$1{,}200 = \$300$

We also know that $I = \Pr t$ with $P = \$1{,}200$ and

$t = 9/12 = 0.75$

Therefore $\$300 = (\$1{,}200)\, r\, (0.75)$ or

$$r = \frac{300}{(\$1{,}200)(0.75)} \approx 33.33\%$$

Example: (Simple Interest Investment)

You wish to save $1,800 to purchase a brand new computer in 14 months. Your local bank offers certificate of deposit that pays a simple interest rate of 2.7%. How much must you invest in a CD from your local bank to come up with $1,800 in 14 months? When necessary, round off your answers to the nearest cent.

Solution:

Future Value is given by $A = P(1 + rt)$ with

$A = \$1,800, r = 0.027,$ and $t = 14/12.$

$\$1,800 = P(1 + (0.027)(14/12))$

$\$1,800 = P(1 + 0.0315)$

$\$1,800 = P(1.0315)$

$P = \dfrac{\$1,800}{1.0315} \approx \$1,745.03$

Another approach would be to use the present value

formula $P = \dfrac{A}{1 + rt}$ directly.

$P = \dfrac{\$1,800}{1 + (0.027)(14/12)} = \dfrac{\$1,800}{1 + 0.0315} = \dfrac{\$1,800}{1.0315} \approx \$1,745.03$

Calculating Compound Interest

Compound interest is calculated based on an exponential growth pattern the amount of interest for compound interest is calculated based on the amount in the account at the beginning of the period, and it continues to grow after each period of time.

If you invest P dollars at an annual interest rate of r, the future value A for the compound interest after t years is calculate by:

$$A = P(1 + r).$$

If the money is compounded n times a year, then the future value after t years is given by:

$$A = P\left(1 + \frac{r}{n}\right)^{nt}.$$

For example, if the money is compounded quarterly, then $n = 4$. P is known as the principal amount or the present value.

Example: (Compound Interest Future Value)
You deposit $10,000 at your local bank at an annual interest rate of 2.5% for two years. Compute the future value of your deposit if the interest is compounded:
    a) annually
    b) semi-annually
    c) quarterly
    d) monthly
    e) continuously

Round off the dollar amount to the nearest cent.

Solution:
The future value of a deposit after t years can be calculated using the formula $A = P\left(1 + \frac{r}{n}\right)^{nt}$, where the present value $P = \$10,000$, n is the number of compounding per year, and r is the interest rate. Approximate your answers to the nearest cent.

a) $A = P\left(1+\dfrac{r}{n}\right)^{nt} = (\$10{,}000)\left(1+\dfrac{0.025}{1}\right)^{(1)(2)} = \$10{,}506.25$

b) $A = P\left(1+\dfrac{r}{n}\right)^{nt} = (\$10{,}000)\left(1+\dfrac{0.025}{2}\right)^{(2)(2)} = \$10{,}509.45$

c) $A = P\left(1+\dfrac{r}{n}\right)^{nt} = (\$10{,}000)\left(1+\dfrac{0.025}{4}\right)^{(4)(2)} = \$10{,}511.08$

d) $A = P\left(1+\dfrac{r}{n}\right)^{nt} = (\$10{,}000)\left(1+\dfrac{0.025}{12}\right)^{(12)(2)} = \$10{,}512.16$

e) The formula for continuous compounding is given by $A = Pe^{rt}$.

$$A = Pe^{rt} = (\$10{,}000)e^{(0.025)(2)} = \$10{,}512.71$$

Example: (Compound Interest Present Value)

Your local bank offers a saving account that earns 3% interest compounded monthly. How much money must you deposit in this account so that it will accumulate to 1$10,000 in five years? When necessary, round off your answers to the nearest cent.

Solution:

Since this is an example of compound interest, you can plug in the numbers in the formula

$A = P\left(1 + \dfrac{r}{n}\right)^{nt}$ and then solve for the present value P.

$$A = P\left(1 + \dfrac{r}{n}\right)^{nt}$$

$$\$10{,}000 = P\left(1 + \dfrac{0.03}{12}\right)^{(5)(12)}$$

$$\$10{,}000 = P(1.0025)^{60}$$

$$P = \dfrac{\$10{,}000}{1.0025^{60}}$$

$$P = \dfrac{\$10{,}000}{1.161617} \approx \$8{,}608.69$$

Another approach would be to use the present value formula $P = \dfrac{A}{\left(1 + \dfrac{r}{n}\right)^{nt}}$ directly.

$$P = \dfrac{A}{\left(1 + \dfrac{r}{n}\right)^{nt}} = \dfrac{\$10{,}000}{\left(1 + \dfrac{0.03}{12}\right)^{(5)(12)}} = \dfrac{\$10{,}000}{1.0025^{60}}$$

$$= \dfrac{\$10{,}000}{1.161617} \approx \$8{,}608.69$$

The **effective annual yield** or **effective rate** Y is the simple interest rate that yields the same future value in an account at the end of one year as when the money is compounded at a rate r for n periods of time in one year. The formula for the effective annual yield Y is given by the following.

$$Y = \left(1 + \frac{r}{n}\right)^n - 1$$

The annual interest rate r is sometimes referred to as the nominal interest rate.

Example: (Effective Annual Yield)
You deposit $10,000 in an account that pays 3.5% interest compounded quarterly.
  a) Determine the effective annual rate.
  b) Find the future value after one year. Round off your answers to the nearest cent.

Solution:
a) The effective yield is defined to be the simple interest rate that results in the same amount of money in the account as the amount produced by the compound interest. The formula for the effective annual yield is given by

$$Y = \left(1 + \frac{r}{n}\right)^n - 1.$$

Therefore we have:

$$Y = \left(1 + \frac{r}{n}\right)^n - 1 = \left(1 + \frac{0.035}{4}\right)^4 - 1 = (1.00875)^4 \approx 3.55\%$$

As expected, the effective rate is slightly higher than the compound rate.

68

b) Using the compound interest rate formula to obtain the future value, we get:

$$A = P\left(1+\frac{r}{n}\right)^{nt} = (\$10{,}000)\left(1+\frac{0.035}{4}\right)^{(4)(1)}$$

$$= (\$10{,}000)(1.00875)^4 = \$10{,}354.62$$

Using the simple interest rate formula, you should get an answer very close to the above.

Annuity is a financial plan that is characterized by equal payments made at equal time periods. The **value** A **of an annuity** is the sum of all the deposits and the total interest paid. If P dollars is deposited at the end of each year for an annuity that is compounded once a year at the annual rate of r, the future value of the annuity after t years is given by

$$A = \frac{P[(1+r)^t - 1]}{r}.$$

Note: We recommend that you round off your answer at the very end. Otherwise, your final answer may be slightly different than the actual answer.

Example: (Annuity Compounded Once a Year)
Suppose when you are 30 years old, you decide to open a saving account for your new born son. You deposit $5,000 into the account at the end of each year for 20 years. If the interest is compounded annually at the rate of 5%, how much money will there be in the account when your son reaches the age of

20? What is the interest after 20 years? Round off your answers to the nearest cent.

Solution:
If P dollars is deposited at the end of each year for an annuity that is compounded once a year at the annual rate of r, the future value after t years is given by

$$A = \frac{P[(1+r)^t - 1]}{r}.$$

Therefore, we have:

$$A = \frac{P[(1+r)^t - 1]}{r}$$

$$= \frac{\$5,000[(1+0.05)^{20} - 1]}{0.05}$$

$$= \frac{\$5,000[(1.05)^{20} - 1]}{0.05}$$

$$\approx \frac{\$5,000[2.6532977 - 1]}{0.05}$$

$$= \frac{\$5,000[1.6532977]}{0.05}$$

$$= \frac{\$5,000[1.6532977]}{0.05}$$

$$\approx \$165,329.77$$

Interest = Future Value – Total Deposits
$$= \$165,329.77 - (\$5,000)(20)$$
$$= 165,329.77 - \$100,000 = \$65,329.77$$

If you deposit P dollars at the end of each period of time for an annuity that is compounded n times a year at the annual rate of r, the future value of the annuity after t years is given by:

$$A = \frac{P[(1+\frac{r}{n})^{nt} - 1]}{\left(\frac{r}{n}\right)}.$$

Example: (Annuity Compounded n Times a Year)
Suppose that in the previous example you deposit $500 at the end of each month at the annual rate of 5% for 20 years. How much money will there be in the account when your son reaches the age of 20? What is the interest after 20 years? Round off your answers to the nearest cent.

Solution:

If P dollars is deposited at the end of each period for an annuity that is compounded n times a year at the annual rate of r, the future value of the annuity after t years is given by:

$$A = \frac{P[(1+\frac{r}{n})^{nt} - 1]}{\left(\frac{r}{n}\right)}.$$

Therefore, we have:

$$A = \frac{P[(1+\frac{r}{n})^{nt} - 1]}{\left(\frac{r}{n}\right)}$$

$$= \frac{\$500[(1+\frac{0.05}{12})^{(12)(20)} - 1]}{\left(\frac{0.05}{12}\right)}$$

$$= \frac{\$500[(1+0.0041667)^{240} - 1]}{0.0041667}$$

$$= \frac{\$500[(1.0041667)^{240} - 1]}{0.0041667}$$

$$= \frac{\$500[2.71266190 - 1]}{0.0041667}$$

$$= \frac{\$500[1.71266190]}{0.0041667}$$

$$\approx \$205,517.78$$

Note: If you plug in the entire expression in your calculator and round off at the end, you should get $205,516.83.

Interest = Future Value − Total Deposits
$$= \$205,516.83 - (\$500)(12)(20)$$
$$= \$205,516.83 - \$120,000 = \$85,516.83$$

The amount of the deposit P dollars that you must make at the end of each period of time for an annuity that is compounded n times a year at the annual rate of r, to achieve a future value of A dollars after t years is given by

$$P = \frac{A\left(\dfrac{r}{n}\right)}{\left[\left(1+\dfrac{r}{n}\right)^{nt} - 1\right]}.$$

Example: (Regular Deposit to Accomplish Financial Goal)

Suppose you wish to save $100,000 for your new born son to be collected by the time he is 20 years old. What amount must you deposit at the end of each month at the annual rate of 3% for 20 years to accomplish this goal? Round off your answers to the nearest cent. How much of the $100,000 comes from deposits and how much comes from interest? Round off your final answers to the nearest cent.

Solution:

The amount of the deposit P dollars that you must make at the end of each period of time for an annuity that is compounded n times a year at the annual rate of r, to achieve a future value of A dollars after t years is given by:

$$P = \dfrac{A\left(\dfrac{r}{n}\right)}{\left[\left(1+\dfrac{r}{n}\right)^{nt} - 1\right]},$$

with $A = \$100{,}000$, $r = 0.03$, $n = 12$, $t = 20$. Therefore, we have:

$$P = \dfrac{(100{,}000)\left(\dfrac{0.03}{12}\right)}{\left[\left(1+\dfrac{0.03}{12}\right)^{(12)(20)} - 1\right]} = \dfrac{250}{\left[(1.0025)^{240} - 1\right]} =$$

$$\dfrac{250}{\left[0.8207549953\right]} = \$304.60$$

Total deposits = $(\$304.60)(12)(20) = \$73{,}104$

Interest = $\$100{,}000 - \$73{,}104 = \$26{,}896$

Bank accounts are insured by the federal government up to the amount of $250,000. As a result, when you invest money in a bank account, you are not taking any chances. Most accounts guarantee an interest rate on your deposit, which will ensure an increase on your investment. This increase is referred to as **return** on your investment. Investing money in a bank account carries little or no risk. The drawback is that the interest rates for the bank accounts are low. There are other kinds of investments that are riskier. You may end up collecting higher interest on your investment, but you may also end up losing part or even all of your deposit. Examples of risk and reward investments are **stocks**, **bonds**, and **mutual funds**. When you purchase a company's stock, you purchase a share of the ownership of the company. For example, if a company issues 100,000 shares and you purchase 1,000 shares, then you own 1,000/100,000 or 1/100 or 0.01 or 1% ownership of the company. For many reasons, a company may issue bonds. Bonds are much less riskier than stocks, but they also offer less return than stocks. When you purchase a bond, you are not buying a share of the company, but instead you are lending money to the company. If the company goes bankrupt, the bondholders will be the first to claim the company's assets. When an investor is unsure of which bond or stock to invest in, the investor may turn to a professional investor for advice. Professional investors manage mutual funds, which consist of a group of selective bonds and stocks.

Information about stocks can be found in daily newspapers and on internet. Example below shows you how to read a daily stock table.

Example: (Stocks)
The table below represents the stock table for a new brand of soda called Fizzy from the TQ Company. Use the stock table to answer questions. When necessary, round off your answers to the nearest cent.

52-Week High	52-Week Low	Stock	SYM	Div	Yld%
58.28	34.56	TQ	Fizzy	.49	1.1%

PE	Vol 100s	Hi	Lo	Close	Net Chg
28	50,320	47.44	40.26	43.29	-1.08

a) What were the high and low prices for a share in the past 52 weeks?

b) If you owned 500 shares of the TQ stock in the last year, how much dividend did you earn?

c) What is the annual return for the dividends alone? How does this compare with a bank offering a 2% simple interest rate?

d) How many shares of the company were traded yesterday?

e) What were the high and low prices for a share of the company yesterday?

f) What was the price at which the share was traded when the stock trade closed yesterday?

g) What was the change in price for a share of the company from the time the exchange closed two days ago to the time the exchange closed yesterday?

h) Use the formula Annual earnings per share = (Yesterday's closing price per share)/(PE ratio) and interpret your result.

Solution:

a) 52-week high = $58.28 and 52-week low = $34.56

b) The price of 1 dividend for the company is listed as $0.49. If you had 500 shares, then you received (500)($0.49) = $245.00.

c) The annual return rate for the dividend alone is given under the column heading Yld%. Therefore, the annual return is 1.1% which is much lower than a bank offering 2% simple interest rate. However, the stock shares for the company may increase in value, making the stock a better investment than the bank account.

d) Volume in hundreds column shows 50,320. Therefore, the number of shares traded yesterday was (50,320)(100) = 5,032,000.

e) The high and low share prices yesterday were high = $47.44 and low = $40.26.

f) At the closing time, a share of the company was worth $43.29.

g) The change in price for a share of the company from the time the exchange closed two days ago to the time the exchange closed yesterday can be found under the column heading "Net Chg" of - $1.08.

h) Annual earnings per share = (Yesterday's closing price per share)/(PE ratio) = $43.29/28 $\approx$ $1.55. The PE ratio of 28 indicates that the yesterday's closing price per share of $43.29 is 28 times greater than the annual earnings per share of approximately $1.55.

Explorations

In The News: Financial Facts

Check a national newspaper, to find stock information. Choose 1 stock and identify each component in the stock table by using the definitions you have learned in this class.

DON'T FORGET TO REFERENCE! See *Section 1* for an example of how.

Writing Across the Curriculum

1. Include copies of web pages from two financial web sites. Write a paragraph describing what you discovered at each website. Write a summary of what you learned overall.

2. Read a self-help book about investment, financial planning, purchasing a home, or other financial topic. Write a report summarizing the most important aspects of the book.

3. Research a company of interest to you. Write a report summarizing basic information about the company and why you would or would not invest in the company.

Chapter 6: Purchasing a Home

At one time or another in our lifetime we all encounter difficult financial decisions involving loans most importantly purchasing a house. In this chapter we are going to concentrate on various practical examples of installment loan options such as calculations of home and car purchases. Once you decide on taking out a loan, you will need to research and seek out the best deal that works for you. You will probably approach several banks or other financial establishments and review their loan options. Perhaps the most important factors that come into consideration are the interest rate on the loan, the monthly payment, and how long it would take for you to pay the loan off. Naturally you would expect the shorter term loan options to carry a lower interest rate and yet higher monthly payments. In the end, you have to decide what financial plan works best for you.

The regular fixed payment amount or PMT needed to repay the loan amount is given by the formula:

$$PMT = \frac{P\left(\dfrac{r}{n}\right)}{\left[1 - \left(1 + \dfrac{r}{n}\right)^{-nt}\right]},$$

where $P =$ the loan amount, $r =$ the annual interest rate, $n =$ the number of payments per year, and $t =$ the number of years to pay off the loan.

Example: (Comparing Loan Options)

Suppose you plan to borrow $15,000 from your local bank to remodel your house. The bank offers you two options:

Option A: 4 year plan at 4.7% annual rate
Option B: 6 year plan at 5% annual rate

Calculate the monthly payment and the total interest for each option. Then create a table to compare the two options. Round off the dollar amount to the nearest cent.

Solution:

The regular payment amount or PMT needed to repay the loan amount can be calculated using the formula:

$$PMT = \frac{P\left(\dfrac{r}{n}\right)}{\left[1-\left(1+\dfrac{r}{n}\right)^{-nt}\right]}.$$

For option A, we have:

$$PMT = \frac{P\left(\dfrac{r}{n}\right)}{\left[1-\left(1+\dfrac{r}{n}\right)^{-nt}\right]} \text{ with}$$

$P = \$15,000, r = 0.047, n = 12, t = 4.$

$$PMT = \frac{(\$15{,}000)\left(\dfrac{0.047}{12}\right)}{\left[1 - \left(1 + \dfrac{0.047}{12}\right)^{-(12)(4)}\right]} = \frac{(\$15{,}000)(0.0039167)}{1 - (1.0039167)^{-48}}$$

$$= \frac{58.75}{1 - 0.8289177} = \frac{58.75}{0.1710823}$$

$$\approx \$343.40$$

The number of payments over 4 years = $(12)(4) = 48$

Total Payments over 4 years
= $(\$343.40)(48) = \$16{,}483.20$

Total amount of interest paid = Total payment − Loan amount = $\$16{,}483.20 - \$15{,}000 = \$1{,}483.20$

For option B, we have:

$$PMT = \frac{P\left(\dfrac{r}{n}\right)}{\left[1 - \left(1 + \dfrac{r}{n}\right)^{-nt}\right]} \text{ with}$$

$P = \$15{,}000,\ r = 0.05,\ n = 12,\ t = 6.$

$$PMT = \frac{(\$15,000)\left(\dfrac{0.05}{12}\right)}{\left[1-\left(1+\dfrac{0.05}{12}\right)^{-(12)(6)}\right]} = \frac{(\$15,000)(0.003333)}{1-(1.0041667)^{-72}}$$

$$= \frac{62.5}{1-0.7412783} = \frac{62.5}{0.2587217}$$

$$\approx \$241.57$$

The number of payments over 6 years = $(12)(6) = 72$

Total Payments over 6 years =
$(\$241.57)(72) = \$17,393.04$

Total amount of interest paid
$\approx \$17,393.04 - \$15,000 = \$2,393.04$

$15,000 Loan	Monthly Payment	Total Interest
Option A: 4 year loan at 4.7%	$343.40	$1,483.20
Option B: 6 year loan at 5%	$241.57	$2,393.04

Next we are going to look at an example of buying a car. In this example, you are given two loan options. Option A offers a longer term loan option with a higher interest rate for a new car, while option B offers a shorter term loan option with lower interest rate for a used car.

Example: (Buying Car – Comparing Loan Options)
Mr. and Mrs. Freeze wish to purchase a car as a graduation present for their 18 year old son. After a month of search, they narrow down the search to the following options.

Option A: a new car that costs $18,000 and can be financed with a 7-year loan at 7%.

Option B: a used car that costs $12,000 and can be financed with a 5-year loan at 5%.

Calculate the monthly payment and the total interest paid for each option. Then prepare a table to compare the two options. Round off your answers to the nearest cent.

Solution:
Option A: (new car paid in 7 years)

Monthly payment:

$$PMT = \frac{P\left(\dfrac{r}{n}\right)}{\left[1-\left(1+\dfrac{r}{n}\right)^{-nt}\right]} = \frac{(\$18,000)\left(\dfrac{0.07}{12}\right)}{\left[1-\left(1+\dfrac{0.07}{12}\right)^{-(12)(7)}\right]} \approx \$271.67$$

Total Interest paid = Total monthly payments – Car price = $(12)(7)(\$271.67) - \$18000 = \$4,820.28$

Option B: (used car paid in 5 years)

Monthly payment:

$$PMT = \frac{P\left(\dfrac{r}{n}\right)}{\left[1-\left(1+\dfrac{r}{n}\right)^{-nt}\right]} = \frac{(\$12{,}000)\left(\dfrac{0.05}{12}\right)}{\left[1-\left(1+\dfrac{0.05}{12}\right)^{-(12)(5)}\right]} \approx \$226.45$$

Total interest paid = Total monthly payments − Car price = $(12)(5)(\$226.45) - \$12{,}000 = \$1{,}587$

	Monthly Payment	Total Interest Paid
Option A: new car 7-year plan at 7%	$271.67	$4,820.28
Option A: used car 5-year plan at 5%	$226.45	$1,587.00

Credit cards are easy to use and you will get to enjoy your purchase immediately. Unfortunately, credit card loans are subject to very high interest rates, and the interest on your unpaid balance can add up to a substantial amount in a short time.

Example: (Credit cards)
Suppose you have a balance of $10,000 on a credit card with an annual interest rate of 12%. You decide to pay off the credit card balance by making regular

monthly payments over a three year period. Calculate the monthly payments and the total interest paid over the three year period. Assume that you will not be making any further purchases on the credit card during the three year period. Round off dollar amount to the nearest cent.

Solution:
To calculate the regular monthly payment, we use the formula:

$$PMT = \frac{P\left(\dfrac{r}{n}\right)}{\left[1-\left(1+\dfrac{r}{n}\right)^{-nt}\right]},$$

with $P = \$10,000$, $r = 0.12$, $n = 12$, and $t = 3$, the monthly payment will be:

$$PMT = \frac{(\$10,000)\left(\dfrac{0.12}{12}\right)}{\left[1-\left(1+\dfrac{0.12}{12}\right)^{-(12)(3)}\right]} = \frac{100}{\left[1-(1.01)^{-36}\right]} = \frac{100}{0.30108} \approx \$332.14$$

Total interest paid = Total monthly payments − Amount of loan

$= (12)(3)(\$332.14) - \$10,00 = \$11,952 - \$10,000 = \$1,957.04$

85

Most credit card companies use the simple interest formula $I = Prt$ to calculate the monthly interest, where $P =$ the average daily balance, $r =$ the interest rate, and $t = 1$ month. The average daily balance is calculated by adding the unpaid balances for each day of the billing cycle and then dividing the sum by the number of days in the cycle. Therefore, the average daily balance is given by:

$$\frac{Sum\ of\ the\ unpaid\ balances\ for\ each\ day\ in\ the\ billing\ cyle}{Number\ of\ days\ in\ the\ cycle}.$$

Example below shows how you can calculate the average daily balance, the interest, the balance due, and the minimum monthly payment for a credit card.

Example: (Credit cards interest, balance due, minimum monthly payment)
Your credit card company uses the average daily balance to calculate the interest with a monthly interest rate of 1.4%. Furthermore, the company requires a minimum monthly payment of $25 if your balance due at the end of the billing cycle is less than $450, or a minimum payment of 1/20 of your balance if your balance is over $450. Table below shows the transactions that have occurred on your credit card during the billing period June 1-June 30. Round off the dollar amounts to the nearest cent.

Transaction Description	Transaction Amount
Previous balance,$5,550	
June 1 Billing date	
June 7 Payment	$1,000 credit
June 10 Charge: Grocery shopping	$150
June 13 Charge: Amazon.com	$420
June 15 Charge: Computer accessory	$250
June 20 Charge: Grocery shopping	$135
June 27 Charge: Clothing	$120
June 30 End of billing cycle	
Payment due date: July 10	

a) Calculate the average daily balance for the billing cycle.
b) Calculate the interest to be paid on the next billing date, July 1.
c) Calculate the balance due on the next billing date, July 1.
d) Find the minimum payment on July 10.

Solution:

a) First we create a table with the beginning date of the billing cycle, each transaction date, and the amount of each unpaid balance.

Date	Unpaid Balance
June 1	$5,550
June 7	$5,550 - $1,000 = $4,550
June 10	$4,550 + $150 = $4,700
June 13	$4,700 + $420 = $5,120
June 15	$5,120 + $250 = $5,370
June 20	$5,640 + $135 = $5,505
June 27	$5,775 + $120 = $5,625

Next we create a new table which includes four columns. One column for the dates, one column for the unpaid balance, one column for the number of days at each unpaid balance, and finally the last column, where we calculate the product of each unpaid balance and the number of days at each unpaid balance.

Date	Unpaid Balance	Number of Days At each Unpaid Balance	Unpaid Balance Times Number of Days
June 1	$5,550	6	($5,550)(6)=$33,300
June 7	$4,550	3	($4,550)(3)=$13,650
June 10	$4,700	3	($4,700)(3)=$14,100
June 13	$5,120	2	($5,120)(2)=$10,240
June 15	$5,370	5	($5,370)(5)=$26,850
June 20	$5,505	7	($5,505)(7)=$38,535
June 27	$5,625	4	($5,625)(4)=$22,500

Total: 30   Total: $159,175

Note: The Number of Days for the last unpaid balance = $30 - (6+3+3+2+5+7) = 30 - 26 = 4$

The average daily balance:

$$\frac{Sum \ of \ the \ unpaid \ balances \ for \ each \ day \ in \ the \ billing \ cyle}{Number \ of \ days \ in \ the \ cycle}$$

$$= \frac{\$159,175}{30} = \$5,305.83$$

b) The interest for the billing cycle can be calculate using the simple interest rate formula $I = P \ rt$, where $P = \$5,305.83$, $r = 0.014$ per month, and $t = 1$ month.
$$I = P \ rt = (\$5,305.83)(0.014)(1) = \$74.28$$

c) The balance due on the next billing date, July 1, is the unpaid balance $5,625 plus the interest of $74.28, or: $5,625 + $74.28 = $5,699.28

d) Since the balance due $5,699.28 exceeds $450, the minimum monthly payment due by July 10 is:
$$\frac{balance \ due}{20} = \frac{\$5,699.28}{20} = \$284.96$$

Mortgage is a long term installment loan arrangement for the purpose of purchasing a home. Typically the mortgage installment plans are made over 15 or 30 year terms. Many sellers require a down payment, which is a percentage of the sale price of the home. The monthly payment can be calculated using the formula

$$PMT = \frac{P\left(\dfrac{r}{n}\right)}{\left[1 - \left(1 + \dfrac{r}{n}\right)^{-nt}\right]},$$

where $P =$ the loan amount, $r =$ the annual interest rate, $n =$ the number of payments per year, and $t =$ the number of years to pay off the home loan.

Example: (Home Mortgage)
The price of a house is $150,000. The bank requires a 5% down payment and two points at the time of closing. You are offered two mortgage options:

Option A: 30 year fixed loan at a rate of 5% per year
Option B: 15 year fixed loan at a rate of 4% per year

Calculate the down payment, the amount of mortgage, the closing costs for two points, the monthly payment and the total interest paid for each option. Then prepare a table to compare the two options. Round off your answers to the nearest cent.

Solution:
Option A:
Down Payment = $(0.05)(\$150,000) = \$7,500$

The amount of mortgage = $150,000 - $7,500 = $142,500

Closing costs for two points on the mortgage =
$0.02 \times \$142,500 = \$2,850$

Monthly payment:

$$PMT = \frac{P\left(\dfrac{r}{n}\right)}{\left[1-\left(1+\dfrac{r}{n}\right)^{-nt}\right]} = \frac{(\$142,500)\left(\dfrac{0.05}{12}\right)}{\left[1-\left(1+\dfrac{0.05}{12}\right)^{-(12)(30)}\right]} \approx \$764.97$$

Total Interest paid = Total monthly payments −
Mortgage =
$(12)(30)(\$764.97) - \$142,500 = \$275,389.20 - \$142,500$
$= \$132,889.20$

Option B:

Down Payment = $(0.05)(\$150,000) = \$7,500$

The amount of mortgage = $150,000 - $7,500 = $142,500

Closing costs for two points on the mortgage =
$0.02 \times \$142,500 = \$2,850$

Monthly payment:

$$PMT = \frac{P\left(\dfrac{r}{n}\right)}{\left[1-\left(1+\dfrac{r}{n}\right)^{-nt}\right]} = \frac{(\$142,500)\left(\dfrac{0.04}{12}\right)}{\left[1-\left(1+\dfrac{0.04}{12}\right)^{-(12)(15)}\right]} \approx \$1,054.06$$

Total interest paid = Total monthly payments − Mortgage =

$$(12)(15)(\$1,054.06) - \$142,500 = \$189,730.80 - \$142,500 = \$47,230.80$$

$142,500 Mortgage	Monthly Payment	Total Interest Paid
Option A: 30 years fixed at 5%	$764.97	$132,889.20
Option A: 15 years fixed at 4%	$1,054.06	$47,230.80

A loan amortization schedule shows the interest payment, the principal payment, and the balance after each periodic payment. An amortization table is an excellent tool to keep track of mortgage. Example below shows you how to create an amortization table.

Example: (Partial Loan Amortization Table)
In the earlier example, after the down payment, a $150,000 home was financed with a $142,500 30 year fixed rate mortgage at 5%. The monthly payment was calculated to be approximately $764.97. Use the given information to create a loan amortization schedule for the first three months of the mortgage. Round off the dollar amount to the nearest cent.

Solution:
We start by calculating the interest, the principal payment, and the loan balance after the first payment.

Interest for the first month =
$$P\,rt = (\$142,500)(0.05)(1/12) = \$593.75$$

Principal payment = Monthly payment − Interest =
$\$764.97 - \$593.75 = \$171.22$

Loan Balance = Principal balance − Principal
payment = $\$142,500 - \$171.22 = \$142,328.78$

We include the information in the first row of the table
and then move on to the calculations for the second
payment.

Payment Number	Interest payment	Principal Payment	Loan Balance
1	$593.75	$171.22	$142,328.78
2			
3			

We calculate the interest, the principal payment, and
the loan balance after the second payment.

Interest for the second month
$= P\,rt = (\$142,328.78)(0.05)(1/12) = \$593.04$

Principal payment = Monthly payment − Interest
$= \$764.97 - \$593.04 = \$171.93$

Loan Balance = Principal balance − Principal
payment = $\$142,328.78 - \$171.93 = \$142,156.85$

We include the information in the second row of the
table and then move on to the calculations for the
third payment.

Payment Number	Interest payment	Principal Payment	Loan Balance
1	$593.75	$171.21	$142,328.78
2	$593.04	$171.93	$142,156.85
3			

We calculate the interest, the principal payment, and the loan balance after the second payment.

Interest for the second month
$= P\, rt = (\$142,156.85)(0.05)(1/12) = \$592.32$

Principal payment = Monthly payment − Interest
$= \$764.97 - \$592.32 = \$172.65$

Loan Balance = Principal balance − Principal payment
$= \$142,156.85 - \$172.65 = \$141,984.20$

We include the information in the second row of the table.

Payment Number	Interest payment	Principal Payment	Loan Balance
1	$593.75	$171.21	$142,328.78
2	$593.04	$171.93	$142,156.85
3	$592.32	$172.65	$141,984.20

A loan may be taken for many different reasons. We are going to conclude this chapter with an example of a loan taken to purchase new furniture.

Example: (Buying new furniture)

You decide to buy new furniture for your house. After visiting several furniture stores, you decide to go with the following offer.

New furniture cost $4,200 that can be financed with a 3-year loan at 7.2%. Calculate the monthly payment and the total interest paid over the three user period.

Solution:

Monthly payment:

$$PMT = \frac{P\left(\dfrac{r}{n}\right)}{\left[1-\left(1+\dfrac{r}{n}\right)^{-nt}\right]} = \frac{(\$4,200)\left(\dfrac{0.072}{12}\right)}{\left[1-\left(1+\dfrac{0.072}{12}\right)^{-(12)(3)}\right]} \approx \$130.07$$

Total Interest paid = Total monthly payments – Price of furniture = $(12)(3)(\$130.07) - \$4,200 = \$482.52$

Explorations

Personal Choices

God blesses us with an abundance of resources. We foster a spirit of service to employ our resources to university and community development. We must be resourceful. We must optimize and apply all of the resources of our community. Part of optimizing and applying all available resources includes making wise decisions with our personal resources. By using our personal resources wisely, we gain knowledge and understanding that enables us to be good stewards when using community resources. Wise choices also allow our resources to prosper with the potential that they may be applied to the needs in our community.

Activity 1

Find a credit card bill (remove all identifying information) that contains an interest charge. Show what the fees and charges would be for the bill using the Average Daily Balance Method, the Unpaid Balance Method, and the Previous Balance Method.

Activity 2

Research to find a credit card that you think would be the best choice for someone who charges regularly, but pays off the bill every month. Find another credit card that you think would be the best choice for someone who makes minimum payments rather than paying off their charges each month. Write an explanation for each of your choices.

1. Should there be a minimum age limit for obtaining a credit card if we respect all people?

2. If Jane is carrying $10,000 in debt on one credit card, should she be allowed to get another credit card?

3. If Jane declares bankruptcy all her debts are "written off" (that means she doesn't have to pay). Is that fair? Consider the value of Respect in your response.

4. How could the value of Respect have helped our country to avoid debt?

Team Writing

Use newspapers, magazines, web pages, or other resources to gather current information as needed. Choose one of the following situations to write a report describing the pros and cons of each situation from the perspective of personal resources. The report should include all details that would be important in making a decision such as interest rates, anticipated fees, intrinsic values, cost analysis, personal lifestyle choices, etc. The report should be a clear summary, with sources referenced appropriately.

a. Buying a home in your area versus renting.

b. Purchasing a new car versus leasing a car.

c. Purchasing new furniture using cash versus credit.

d. Investing $10,000 in a CD versus stocks versus bonds versus mutual funds.

To Retire or not to Retire (*Teams have one member do each activity then compare to see what is most important in investment decisions. Individuals choose one activity to complete*)

1. Research to find the average return over the last five or ten years on a mutual fund of your choice. Use that average return to see what your investment account would look like if you had invested $10,000 for 5, 10, 15, 20, 25, 30, 35 and 40 years compounded quarterly. Show your result numerically and graphically. Is there a point at which your investment appears to grow more quickly? What can you conclude about the effect of time on retirement investment?

2. Research to find the average return over the last five or ten years on a mutual fund of your choice. Use that average return to see what your investment account would look like if you had invested $5,000, $10,000, $15,000 and $20,000 for 35 years compounded quarterly. Show your result numerically and graphically. What can you

conclude about the effect of initial investment on retirement investment?

3. Research to find the average return over the last five or ten years on a mutual fund of your choice. Use that average return to see what your investment account would look like if you had invested $10,000 at that rate of return, at 2% less and at 2% more for 35 years compounded quarterly. Show your result numerically and graphically. What can you conclude about the effect of rate of return on retirement investment?

Discussion

1. Does government have a responsibility to ensure that its citizens prepare for retirement?

2. Is society obligated to care for the elderly who have been unable, unwilling or unfortunate in their preparation for retirement?

Chapter 7: Counting Principles

You probably already thought you knew how to count. Counting falls within the mathematical field of Combinatorics, so we will expand on traditional counting in order to solve counting problems. The methods of counting covered in this chapter are permutations and combinations.

In order to be able to use the formulas for permutations and combinations, let's first look at factorial notation. **Factorial notation** is given by an exclamation mark after a constant. It means to multiply that number by every natural number less than it. For example,

$3! = 3*2*1 = 6$

$7! = 7*6*5*4*3*2*1 = 5,040$

In general the formula looks like

$n! = n*(n-1)*(n-2)*... (3)*(2)*(1)$

Factorials can be simplified in fractions because the numbers are multiplied together. As a result, if you have the same factorial in the numerator as the denominator, that factorial can be canceled. It is the same as canceling each of the factors within the factorial. For example, if you have $\frac{14!}{(12)!}$ you can break the numerator down into 14*13*12! since that is still the product of all the natural numbers below and including 14. Now your fraction becomes: $\frac{14 * 13 * 12!}{(12)!}$ and you can cancel the 12! from the numerator and denominator because it is the same as having the common factors of 12, 11, 10, 9, 8, 7, 6, 5, 4, 3, 2 and 1. However, you can cancel all 12 factors in one step by canceling the factorial to be left with 14*13 = 182. This is especially helpful when the factorials are too large for your calculator to handle.

Fundamental Counting Principle

The Fundamental Counting Principle states that if you have m items of one type and n items of another type, there are m*n ways you can combine those items. We can use this principle to determine the number of arrangements that can be made in many different situations.

Example:
How many unique phone numbers can be created for each area code?

Solution:
Each phone number has seven digits after the area code. Each digit can be chosen from the numbers 0 through 9 so there are 10 choices for each digit. That means there are 10*10*10*10*10*10*10 = 10^7 = 10,000,000 different phone numbers possible for each area code.

Permutations

Permutations are the number of ways a certain number of objects can be arranged if the order of the arrangement is important, so that each different arrangement gives a different result. In other words, if the order makes a new arrangement, you are working with permutations. An example would include electing officers in a club. If there is a president, secretary, and treasurer, you might be choosing three officers from a club with 20 members. The order is defined by an office, so that if you choose three students but change the order of those students, you have a new arrangement because the students would be filling a different office. The notation for choosing three officers from a club of 20 members looks like $_{20}P_3$. This means the number of permutations of 20 members taken three at a time.

Permutations can be calculated using commands on a graphing calculator, on Excel, or by hand. The formula for the number of permutations of n items taken k at a time is given by:

$$_nP_k = \frac{n!}{(n-k)!}$$

Example:
The 5th grade class at Madeira Fundamental Elementary School has 24 students and must choose a representative and an alternate student government representative. How many different ways can they choose the representative and the alternate?

Solution:
Since there are 24 students two at a time and the order changes whether the student is the representative or alternate, we would calculate the number of permutations of 24 items taken two at a time to have:

$$_{24}P_2 = \frac{24!}{(24-2)!} = \frac{24!}{(22)!} = \frac{24 * 23 * 22!}{(22)!} = 24*23 = 552 \text{ so}$$

there are 552 different ways a representative and an alternate can be chosen from the 24 students in the class.

Example:

How many different teams can you put together in volleyball if you have 12 hitters and five play at a time with your setter?

Solution:

Since each arrangement puts different players next to each other and in different positions on the court, each arrangement of five players is different each time you move a player. As a result, this is the same as the number of permutations of 12 players taken five at a time or

$$_{12}P_5 = \frac{12!}{(12-5)!} = \frac{12!}{(7)!} = \frac{12 * 11 * 10 * 9 * 8 * 7!}{(7)!} =$$

$12 * 11 * 10 * 9 * 8 = 95,040$ different player arrangements on the court. Too bad permutations don't tell you what arrangement will win.

These calculations can be done on many graphing calculators, or on Excel using the following command: =PERMUT(12,5) and then select "enter"

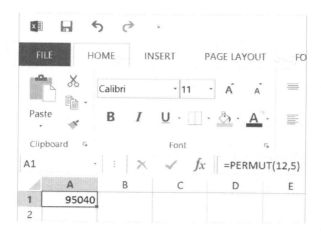

Combinations

Next is combinations and then how to determine which to use. Notice with permutations, each arrangement was different if the order of the objects was changed. With **Combinations**, the order of the objects does not matter, only the objects themselves. Combinations might include if your group wanted to choose three representatives for a committee. The order of the people chosen will not change the committee of individuals. Or when investing, the number of ways to choose five stocks for your portfolio. The order in which you choose them will not matter, only the fact that you chose each particular group of five stocks. Or,

Combinations can also be calculated using commands on a graphing calculator, on Excel, or by hand. The formula for the number of combinations of n items taken k at a time is given by:

$$_nC_k = \frac{n!}{(n-k)!k!}$$

Example:
A flight has 3 open seats but there are 7 people on stand-by. How many different ways can the three people be chosen?

Solution:
Since the 3 people will all get a seat, it does not matter what order they are chosen, so this is a combination problem with 7 items taken 3 at a time.

$$_7C_3 = \frac{7!}{(7-3)!3!} = \frac{7!}{(4)!3!} = \frac{7*6*5*4!}{4!*3*2*1} = \frac{7*6*5}{3*2*1} = 35$$

We can subtract 3 from 7 to get 4. Then 7! can be broken down to 7*6*5*4!. Notice we stopped at 4! because that will cancel in the denominator leaving the 3! as 3*2*1 and the 7*6*5 in the numerator. Simplifying the numerator and denominator we see that a factor of 6 will cancel leaving the total number of arrangements as 35.

Now we have two methods to use to count the number of arrangements of n items taken k at a time. However, that means we need to be able to correctly choose between the methods.

Example:
Let's consider a race where the top three finishers go on to race in the final round. How many different arrangements

of three people can go on to the final round if there are 15 runners in the race?

Solution:
We will first need to determine if this is an application of combinations or permutations. In this case, we need to decide if changing the order gives us the same or a new arrangement. Since changing the order of the finishers will not change the fact that the same three progress to the finals, this is a combinations problem. We are finding the number of combinations of the 15 runners taken 3 at a time.

$$_{15}C_3 = \frac{15!}{(15-3)!3!} = \frac{15!}{(12)!3!} = \frac{15*14*13*12!}{12!*3*2*1} = \frac{15*14*13}{3*2*1} = 455$$

Here we can first subtract 3 from 15 to get 12 factorial in the denominator so we count down the 15! to 12! So we can cancel 12! From the numerator and denominator, leaving 15*14*13 on top and 3*2*1 on the bottom. We can cancel a factor of 6 leaving 455 different arrangements for the combination of three winners. Of course, if we wanted the winners to get a first, second, and third place ribbon, this would not become a problem for permutations because each arrangement would change the ribbons that the runners would receive.

This problem can also be calculated using Excel. The command is =COMBIN(15,3):

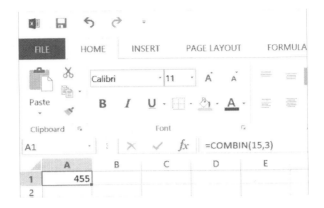

Probability

We will focus on two types of probability in this chapter, Theoretical and Empirical. Theoretical Probability is the probability of something happening in theory. That means that it is the number of times something can occur divided by the total number of possible occurrences when the outcomes are equally likely. Empirical probability is the actual rate at which the outcome occurred in past experience. In both cases we have the basic probability expressed as P(x) meaning the probability that x will occur.

$$\text{Theoretical } P(x) = \frac{\textit{the number of ways x can occur}}{\textit{the total number of possible outcomes}}$$

$$\text{Empirical } P(x) = \frac{\textit{the number of times x did occur}}{\textit{the total number of outcomes}}$$

We can use things like cards or dice to calculate either of these probabilities. For example, let's say a die was rolled 15 times with the following outcomes:

Outcome of roll	1	2	3	4	5	6
Number of outcomes	3	4	2	2	1	3

We could find the probability of rolling a 3 on the die.

$$\text{Theoretical } P(x) = \frac{\textit{the number of ways x can occur}}{\textit{the total number of possible outcomes}}$$

$$= \frac{1}{6}$$

$$\text{Empirical } P(x) = \frac{\textit{the number of times x did occur}}{\textit{the total number of outcomes}} = \frac{2}{15}$$

Notice that the probabilities are different for the same outcome. For Theoretical probability, there is only one way to roll a 3 and six different possible numbers that could come up. For Empirical probability there were a total of 15 rolls and 2 of them were the number three.

There is actually a "Law of Large Numbers" that says that as the sample size gets larger, the Empirical probability will approach the value of the Theoretical probability.

Theoretical Probability:

Some simple examples of Theoretical probability would be determining the probability of selecting a particular card or rolling a particular number on a die. For example the probability of rolling a number less than 3 on a die is the number of outcomes that are less than 3 (two possible outcomes less than 3) divided by the total number of outcomes (6 total numbers on a die) for the probability of P(rolling a number less than 3) = 2/6 = 1/3.

Example:

We can determine theoretical probability for things such as:

a. Find the probability of drawing a heart from a standard deck of cards.

b. Find the probability of having exactly 2 girls out of three children.

Solution:

a. Since there are 52 cards in the standard deck, and four suits. That means 13 of those 52 cards are hearts, so

$$P(\text{heart}) = \frac{13}{52} = \frac{1}{4}$$

b. The possible outcomes of gender for three children are made up by the set of these eight possible outcomes: {ggg, gbb, gbg, bgg, bbg, bgb, gbb, bbb}

$$P(\text{exactly 2 girls}) = \frac{3}{8}$$

The examples can also get more complicated such as examples that use permutations for combinations to determine the total number of outcomes.

Example:

If you selected 3 stocks from a group of 9 that increased in value and 6 that decreased in value this quarter, what is the probability that all three stocks increased in value?

Solution:

To find the probability that all 3 stocks went up in value, we would find the ratio of the number of ways of selecting 3 stocks that went up in value to the number of ways of selecting any 3 stocks. Since the order the stocks are chosen does not matter, it will still be the same 3 stocks, this problem uses combinations to determine the total number of ways to select 3 stocks.

A = selecting 3 stocks that increased in value

P(A) =

Number of ways of selecting 3 stocks from the 9 that increased

Number of ways of selecting 3 stocks from all the stocks

$$P(A) = \frac{_9C_3}{_{15}C_3} = \frac{\dfrac{9!}{(9-3)!3!}}{\dfrac{15!}{(15-3)!3!}} = \frac{\dfrac{9!}{(6)!3!}}{\dfrac{15!}{(12)!3!}} = \frac{\dfrac{9*8*7*6!}{(6)!3!}}{\dfrac{15*14*13*12!}{(12)!3!}} =$$

$$\frac{\dfrac{9*8*7}{3!}}{\dfrac{15*14*13}{3!}}$$

$$= \frac{\dfrac{9*8*7}{3*2*1}}{\dfrac{15*14*13}{3*2*1}} = \frac{84}{455} = \frac{12}{65}$$

Calculating the combinations using technology will save you a lot of steps and time so you could have gone straight to:

$$\frac{{}_9C_3}{{}_{15}C_3} = \frac{84}{455} = \frac{12}{65}$$

Empirical Probability

The second type of probability we will explore is Empirical Probability. It is the probability calculated from previous results, or empirical data. In other words, it is the ratio of the actual number of times a particular outcome occurred compared to the total number of outcomes that occurred.

P(x) =
$$\frac{Number\ of\ outcomes\ that\ satisfy\ the\ condition\ of\ x}{Total\ number\ of\ outcomes\ possible}$$

Notice that since the equation is the same, it is how we find the number of outcomes that is different.

Example:
If someone is in the department of math and sciences, what is the probability they are a math major?

	Men	Women
Math major	18	16
Science major	27	39

Solution:
Since the data is given, you add up the math majors and divide it by the total number of students in the department to get:

$$P(\text{math major}) = \frac{number\ of\ math\ majors}{number\ of\ majors} =$$

$$\frac{18 + 16}{18 + 16 + 27 + 39}$$

$$= \frac{34}{100} = \frac{17}{50}$$

Probabilities can be expressed in fraction, decimal, or percentage form.

Explorations

Writing Across the Curriculum

1. Explain in your own words why empirical probabilities are used in determining premiums for life insurance policies.

2. Explain in your own words, how to find the theoretical probability of an event.

3. How are permutations and the fundamental counting principle the same? How are they different?

4. How are Theoretical and Empirical probability the same? How are they different?

5. What is Combinatorics? Who is an influential mathematician in that field and what has he/she done?

Critical Thinking and Values for Effective Problem Solving:

1. Two playing cards are dealt to you from a well-shuffled deck of 52 cards. If either card is a diamond, or both are diamonds, you win; otherwise, you lose. Determine whether this game favors you, is fair, or favors the dealer. Explain your answer.

2. Should some forms of gambling, such as this one, be legalized? Why or why not? Discuss who is often hurt by gambling and who is often helped.

3. What forms of gambling are legal and what forms are not? Give examples. Should gambling be allowed on

Indian reservations in states where it is not allowed off those reservations? Why or why not?

Team Exercise:

Can people selected at random distinguish Coke from Pepsi? Design an experiment to determine the empirical probability that a person selected at random can select Coke when given samples of both Coke and Pepsi. Describe the experiment and document your results.

Exploring Number Theory

Fall 2013 Headcount:	
Total University Enrollment	16,275
University Campus, Saint Leo, Florida	2,234
Graduate Programs	3,650
Adult Education Center	409
Regional Centers:	7,057
Center for Online Learning	3,162
Online Consortium of Independent Colleg and Universities	172
Saint Leo University Facts and Figures 2013-2014	

1. Based on the data given, what were the odds that a randomly selected student was from a graduate program at SLU in fall 2013? Based on the data, how many SLU students were fully online in all? Choose one of the specific locations and determine the odds that a randomly selected student came from that location. Make sure you can justify your answers.

2. Choose two or more of the specific locations and determine the probability that a randomly selected student attended either of the locations in fall 2013. For example, find the probability that a randomly selected student attended a regional center or the adult education Center?

3. Choose one of the locations and determine the total number of ways that a committee of 3 students could have been chosen to represent that location.

4. Choose one of the locations and determine the total number of ways that a committee consisting of a lead representative, second representative and alternate could have been chosen to represent that location.

Chapter 8: Probability

In the last chapter we found the probability an event would occur. In this chapter we will begin by finding the probability an event will not occur. This concept is called the compliment and is expressed as $P(\bar{x})$ and it is given by $P(\bar{x}) = 1 - P(x)$

Example:
When you draw a card from a standard deck, find the probability that the card is not a five

Solution:
Since there are 52 cards in the deck and four of them are the number 5, that means the probability of drawing a 5 is $\frac{4}{52} = \frac{1}{13}$ so the probability that the card is not a five is 1 - $\frac{1}{13} = \frac{12}{13}$.

Example:
Using the table below to determine the following:
a. The probability that a man in the Department of Mathematics and Sciences majored in Science?

b. The probability that a man in the Department of Mathematics and Sciences did not major in Science?

	Men	Women

Math major	18	16
Science major	27	39

Solution:

a. $P(x) = \dfrac{number\ of\ men\ who\ majored\ in\ science}{total\ number\ of\ men} =$

$\dfrac{27}{18+27} = \dfrac{27}{45} = \dfrac{3}{5}$

b. $P(\bar{x}) = 1 - P(x) = 1 - \dfrac{3}{5} = \dfrac{2}{5}$

"And" and "or" Probabilities

We can also find the probability of more than one event occurring. We might be trying to determine if two events happened together, or if one or more of two events happened at all.

For two events happening together, we multiply their separate probabilities together or for empirical data we find the overlapping cells in the table of data.

For one "or" another probability happening, we add their probabilities in such a way that we do not add any outcome more than one time. If we add totals that contain overlapping cells, then we would need to subtract those overlapping cell values so they are only included once in our total.

Example:

Consider the table below, of students in the Honors Program. We can find several different combined probabilities such as those listed below.

	Freshman	Sophomore	Junior	Senior	Total
Male	66	62	51	35	214
Female	71	68	59	41	239
Total	137	130	110	76	453

a. Find the probability that an honors student is an upper classman.
b. Find the probability that an honors student is female or a sophomore.
c. Find the probability that an honors student is a freshman.
d. Find the probability that an honors student is not a freshman.
e. Find the probability that an honors student is a male senior.
f. Find the probability that an honors student is a senior or male.

Solution:

a. P(upper classman) = $\dfrac{number\ of\ Juniors\ and\ seniors}{number\ of\ honors\ student}$ =
$\dfrac{110 + 76}{453}$ = $\dfrac{186}{453}$ = $\dfrac{62}{151}$

b. P(female or sophomore) =
$$\frac{number\ of\ females\ and\ sophomore\ without\ repeating}{number\ of\ honors\ student}$$

$$= \frac{239 + 130 - 68}{453} = \frac{301}{453}$$

(Total females plus total sophomores, less female sophomores)

$$or = \frac{71 + 68 + 59 + 41 + 62}{453} = \frac{301}{453}$$

(Female freshman plus female sophomores plus female juniors plus female seniors plus male sophomores)

c. P(freshman) = $\frac{number\ of\ freshman}{number\ of\ honors\ student} = \frac{137}{453}$

d. P(not a freshman) = 1 - P(freshman) = $1 - \frac{137}{453} = \frac{316}{453}$

e. P(male senior) = P(male and senior at the same time) =
$$\frac{number\ of\ males\ and\ seniors\ at\ the\ same\ time}{number\ of\ honors\ student} = \frac{35}{453}$$

f. P(senior or male) =
$$\frac{number\ of\ males\ and\ seniors\ without\ repeating}{number\ of\ honors\ student} =$$

$$\frac{214 + 76 - 35}{453} = \frac{325}{453}$$

Example:

119

Given a deck of cards:

a. Find the probability of drawing a 3 or a queen.
b. Find the probability of drawing a 2 and a 6.
c. Find the probability of drawing a 2 and a club
d. Find the probability of drawing a king or a heart.

Solution:

a. P(3 or queen) =

$$\frac{Total\ number\ of\ 3s\ and\ queens\ without\ repeating}{Total\ number\ of\ cards}$$

$$= \frac{4+4}{52} = \frac{8}{52} = \frac{2}{13}$$

b. P(2 and 6) =

$$\frac{Total\ number\ of\ cards\ that\ are\ both\ a\ 2\ and\ a\ 6\ at\ the\ same\ tim}{Total\ number\ of\ cards}$$

$$= \frac{0}{52}$$

c. P(2 and club) =

$$\frac{Total\ number\ of\ cards\ that\ are\ either\ a\ 2\ or\ a\ club\ without\ repe}{Total\ number\ of\ cards}$$

$$= \frac{4+12-1}{52} = \frac{15}{52}$$

d. P(king or heart) =

$$\frac{Total\ number\ of\ cards\ that\ are\ a\ king\ or\ a\ heart\ without\ repeating}{Total\ number\ of\ cards}$$

$$= \frac{4\ kings + 12\ hearts - 1\ king\ of\ hearts}{52} = \frac{15}{52}$$

Conditional probability

We can refine our probability calculations based on specific categories when conditions are given. For example, if we look at the math and science majors example above, if we know the person is a math major, we only have to calculate the probability out of all math majors rather than all students. Conditional probability means you first limit the outcomes to only those that meet the condition, then calculate the probability within that restricted pool of values. The notation for conditional probability is $P(a \mid b)$ or the probability of a given b.

Example:
Use the given data to find the following probabilities.

	Freshman	Sophomore	Junior	Senior	Total
Male	66	62	51	35	214
Female	71	68	59	41	239
Total	137	130	110	76	453

a. The probability that someone is a sophomore given that they are female.
b. The probability that someone is male given that they are a senior.
c. The probability that someone is an upper classman given that they are male.

Solution:
a. P(sophomore | female) =

$$\frac{Total\ number\ of\ females\ who\ are\ sophomores}{Total\ number\ of\ females} = \frac{68}{239}$$

b. P(male | senior) =
$$\frac{Total\ number\ of\ seniors\ who\ are\ males}{Total\ number\ of\ seniors} = \frac{35}{76}$$

c. P(upper classman | male) =
$$\frac{Total\ number\ of\ males\ who\ are\ upper\ classman}{Total\ number\ of\ males} =$$
$$\frac{51 + 35}{214}$$
$$= \frac{86}{214} = \frac{43}{107}$$

Odds

A concept very similar to probability is odds. The difference is that odds are the number of outcomes in favor divided by the number of outcomes against, while probability is the number of outcomes in favor divided by the total number of outcomes. The numerators are the same, but the denominator for odds is the difference between the numerator and denominator of the probability. That means if the probability of an outcome is given by P(x) = $\frac{a}{b}$ then the odds are given by odds in favor of x = $\frac{a}{b-a}$.

The odds against x would be the reciprocal since it would be the number outcomes against divided by the number of outcomes in favor.

Odds can also be expressed as a: (b-a) or a to (b-a)

Example:
The table below shows the number of people who have each type of pet in a certain survey.

	Number
Dog	64
Cat	31
Bird	5
No Pet	30
Total	130

a. Find the odds that a randomly selected person has a dog.
b. Find the odds that a randomly selected person does not have a dog.
c. Find the odds that a randomly selected person has a bird.
d. Find the odds that a randomly selected person has a cat or a bird.

Solution:

a. Odds of a dog $= \dfrac{number\ of\ dogs}{number\ that\ are\ not\ dogs} =$

$\dfrac{64}{31+5+30} = \dfrac{64}{66} = \dfrac{32}{33}$ or 32:33

b. Odds of not a dog $= \dfrac{1}{odds\ of\ a\ dog} = \dfrac{1}{\frac{32}{33}} = \dfrac{33}{32}$

c. Odds of a bird $= \dfrac{number\ of\ birds}{number\ that\ are\ not\ birds} = \dfrac{5}{130} = \dfrac{1}{26}$
or 1:26 or 1 to 26

d. Odds of a cat or a bird $= \dfrac{number\ of\ a\ cat\ or\ a\ bird}{number\ that\ are\ not\ cats\ or\ birds} = \dfrac{31+5}{64+30} = \dfrac{36}{94}$ or
18:47 or 18 to 47

Expected Value:

Many games of chance or bids for a contract are based on expected value. They each have a cost associated with them such as the cost to put together a bid on a project, or the cost to enter a game of chance. They also have a probability of success and a relatively defined pay-out if successful. When you put these concepts together you can determine the expected value or average outcome if you were to participate repeatedly. If you expect to have a positive outcome in the long run, then you might choose to participate, if you expect to lose in the long run, you should probably pass on the bid or game of chance. Casino games always have a negative expected value because that is how the casino makes money, by the overall average going to the casino exceeding the amount they pay out, even though there are occasional people who win more than they spend.

Example:
For the example below, find the expected value for each contract and determine whether you should bid on one, the other, both, or neither contracts. The bid cost is how much it will cost you to make the bid, Payment is the amount you can expect to make if you are awarded the contract. The probability of hire is the probability you will win the bid.

	Bid Cost	Payment	Probability of Hire	Expected Value
Contract A	$6000	$170,000	.25	
Contract B	$15000	$460,000	.15	

Solution:
Expected value of Contract A = $170,000*(.25) − $6000 = $36,500
Expected value of Contract B = $460,000*(.15) - $15,000 = $54,000

The expected value of both contracts is positive so they are worthwhile pursuing.

Explorations

In The News: Financial Facts

Go to a newspaper database and search for "probability". Choose one of the articles from your search results and explain how the term Probability is used in that article. Also explain why the given probability is or is not used correctly in the article.

DON'T FORGET TO REFERENCE! See *Section 1* for an example of how.

Critical Thinking with Probability

Almost anyone in the United States can play the lottery and lottery games come in many kinds. You might be surprised how many people dedicate $50 (or more) a week to playing lotto games. But can the game you choose significantly affect you chances of winning? For the next two exercises you will need to use the internet to find information.

1. Choose a state which runs lotteries.
   a. Calculate your odds of winning if you spend $1 on an entry.
   b. Calculate your odds of winning if you spend $50 on an entry.
   c. Compare these odds to the odds of being in a car accident, plane crash, struck by lightning, or hit by a meteorite. (These numbers *are* out there so search for them!)

2. At the rate of $50/week you will have spent $10,000 in less than 5 years. Take a look at the results of exercise

1 through 3 under Money Talk in the previous section. Given those results and your calculation of the odds, are lotteries a good personal investment?

Discussion:

1. While many states run lotteries, most prohibit individuals and businesses from doing so. Is that ethical?

2. Court cases have held bartenders liable for serving alcohol to someone who is already drunk. Gambling is an addiction like alcohol. Should a state be liable for selling lotto tickets to a gambling addict?

Writing Across the Curriculum

1. Write and explain the formula used to find the expected value of an experiment with two possible outcomes and with three possible outcomes.
2. If the expected value and cost to play are known for a particular game of chance, explain how you can determine the fair price to pay to play that game of chance. Give the formula for determining the fair price to pay to play a particular game of chance with three possible *gross* amounts that can be won.

3. The dealer shuffles five black cards and five red cards and spreads them out on the table face down. You choose two at random. If both cards are red or both cards are black, you win a dollar. Otherwise, you lose a dollar. Determine whether the game favors you, is fair, or favors the dealer. Explain your answer.

4. If events A and B are mutually exclusive, explain why the formula P(A or B) = P(A) + P(B) − P(A and B) can be simplified to P(A or B) = P(A) + P(B).

5. Explain how to determine probabilities when you are given an odds statement.

Made in the USA
Middletown, DE
04 February 2017